川山茶古树 名·录

Records of Sichuan Ancient Camellia Trees

周 利 主编

重庆出版集团 重庆出版社

《川山茶古树名录》编委会

主　编：周　利

副主编：李玲莉、宋春艳

编　委：周　利、李玲莉、宋春艳、尹有惠、谭崇平、何　洁、胡　莹、王　黎、陈　莉

照　片：周　利、李玲莉、王　宇、何　洁

支持单位：重庆市南山植物园管理处、重庆市住房和城乡建设委员会科技与对外合作处

Editorial Committee of Records of Sichuan Ancient Camellia Trees

Editor: Zhou Li

Vice-editor: Li Lingli, Song Chunyan

Member: Zhou Li, Li Lingli, Song Chunyan, Yin Youhui, Tan Chongping, He Jie, Hu Ying, Wang Li, Chen Li

Photo: Zhou Li, Li Lingli, Wang Yu, He Jie

Support unit: Chongqing Nanshan Botanical Garden Administrative Office, Science and Technology and Foreign Cooperation Office of Chongqing Housing and Urban Rural Construction Commission

保护山茶古树
造福子孙后代

贺"川山茶古树名录"付梓

管开云
2022年8月10日

管开云
中国科学院昆明植物研究所博士、研究员、博士生导师
前国际山茶协会主席

> Protect ancient camellia trees.
> Bring benefit to the coming generations.
> Congratulations to the publication of the book 'Records of Sichuan Ancient Camellia Trees'.
>
> Guan Kaiyun
> 10 Aug. 2022

Prof. Guan Kaiyun
PhD, Professor, Kunming Institute of Botany, Chinese Academy of Sciences (CAS)
Former President of International Camellia Society

川山茶古树
Records of Sichuan Ancient Camellia Trees
名·录

序言1

川山茶古树调查，
意义深远，必将大放异彩

获悉重庆南山植物园周利高工要出版一部《川山茶古树名录》，让我写序，倍感荣幸。

四川和重庆一带，是我国乃至世界上山茶属植物分布最为丰富的地区之一，也是我国西南地区山茶古树分布数量最多的区域。据我国山茶属植物分类专家、中山大学教授张宏达先生的学术著作记载，我国四川金沙江一带，包括重庆地区，是红山茶物种的发源地，其山茶古树之多，令世人叹为观止。山茶古树是祖先留给后人的宝贵财富，是活的文物、古董，具有经济、社会、生态、历史文化和遗传学利用价值。

其实，早在2003年，我国就已经开始对野生的山茶物种进行了调查和收集。浙江省金华市国际山茶物种园的成功营建，为野生的山茶古树的保育奠定了基础。2010年，对我国茶花事业做出巨大贡献的游慕贤先生，经多年实地考察，一部震惊国内外茶花界的巨著《中国茶花古树觅踪十年》问世，将我国茶花古树的保育进一步推向高潮。

2019年国际山茶协会(International Camellia Society, ICS)山茶古树保育大会在我国广东省广州市从化区阿婆六茶花谷隆重召开，并首次就山茶界今后对山茶古树保育事宜进行专题讨论，会议期间隆重发布了《阿婆六宣言》。

长江后浪推前浪，《川山茶古树名录》一书，通过科学的调查方法，

对446株山茶古树及后备资源进行了详细调查和记述，其中387株栽培的川茶传统茶花品种古树及后备资源，59株是野生类型的山茶古树及后备资源。这标志着四川和重庆一带的山茶古树资源已经清晰，这无论对我国今后茶花产业的发展，还是对整个茶花界都可谓是一个巨大贡献，由此将进一步推动我国山茶古树的保护和深度开发利用。

这些调查和记述的山茶古树，在它们的历史长河中，都是通过百年甚至几百年与逆境抗争而得以生存下来的，它们的生命力是何其强大！可以设想，用这些山茶古树作为亲本材料，通过杂交途径，获得的杂交后代，其抗性一定会青出于蓝而胜于蓝。从观赏角度看，这些山茶古树，树体通常高大，枝条苍劲有力，树姿优美，开花稠密，是弘扬茶花文化，推动旅游业发展极其宝贵的资源。

祝贺周利高工新书的出版，我坚信，川山茶古树，必将吸引世界同行的目光，它们将大放异彩。

高继银

中国林科院亚热带林业研究所研究员
广东肇庆棕榈谷花园有限公司茶花育种团队首席专家
2022年6月20日

Foreword I

The investigation of Sichuan ancient camellia trees, far-reaching significance, will shine

It is my great honor to know that Mrs. Zhou Li, senior engineer of the Chongqing Nanshan Botanical Garden is publishing a book named *Records of Sichuan Ancient Camellia Trees* and she would like me to write a foreword for her book.

Sichuan and Chongqing are not only among the regions with the most abundant distribution of camellia plants in China and even worldwidely, but also among the regions with the largest distribution of ancient camellia trees in southwest China. According to the recordation of the monographs written by Mr. Zhang Hongda, an expert on camellia classification in China and a professor of Sun Yat-sen University, Jinshajiang area of Sichuan Province, China, including Chongqing area, is the birthplace of *C. japonica*. There are surprisingly large quantities of ancient camellia trees that is astonishing to the world. Ancient camellia trees, precious heritage from our ancestors, are now living cultural relics and antiques that has great potential for utilization in economic, social, ecological, historical, cultural and genetic aspects.

In fact, China has started investigating and collecting wild camellia species as early as 2003. The successful construction of Jinhua International Camellia Garden in Zhejiang province has laid a foundation for the conservation of wild ancient camellia trees. In 2010, Mr. You Muxian, who has made great contributions

to China's camellia career, published a great book *Tracking down Chinese Ancient Camellia Trees for Ten Years* after years of investigation and field trips. This book is astonishing to fields of camellia research nationwidely and worldwidely, thus further pushed the conservation of China's ancient camellia trees to a new climax.

The ICS Ancient Camellia Tree Conservation Conference was held at Apoliu Camellia Valley, Conghua District, Guangzhou city, Guangdong Province, China in 2019. For the first time, a special discussion on the conservation of ancient camellia trees of the camellia field in the future. During the conference, The Declaration of Apoliu was solemnly issued.

As same as the waves of Yangtz River that the new research followed by the old, the book *Records of Sichuan Ancient Camellia Trees* investigated and recorded 446 ancient camellia trees comprehensively through scientific investigation methods, and among which 387 were cultivated traditional ancient camellia trees and 59 were wild ancient camellia trees. The publication of this book indicates that the resources of ancient camellia trees in Sichuan and Chongqing areas have been clear, and is a great contribution to the future development of China's camellia field and the whole camellia field that will further promote the protection and in-depth development and utilization of ancient camellia trees in China.

Those investigated and recorded ancient camellia trees have survived

through a hundred or even several hundred years of struggling in their long history. I'm amazed by how powerful they are! It can be imagined that hybrid offspring crossbreeding using those ancient camellia trees as parents should be better than their parents. From an ornamental point of view, these ancient camellia trees are usually tall and beautiful, with vigorous branches and dense flowers, thus are extremely valuable resources for cultural promotion and the development of tourism.

    Accompanied by the publication of this book, I'm certain that Sichuan ancient camellia trees will become a rising star and attract great attentions of colleagues among the world.

Mr. Gao Jiyin
Professor, Research Institute of Subtropical Forestry, Chinese Acadeny of Forestry (CAF)
Chief Expert, Camellia Breeding Team of Zhaoqing Palm Valley Garden Co. Ltd, Guangdong Province, China
20 June, 2022

序言 2

川渝地区是我国重要的山茶属植物分布区，野生资源十分丰富。据不完全统计，该地区发表的山茶属的物种名称超过 30 个，是西南山茶（*C. pitardii*）、怒江山茶（*C. saluenensis* Stapf ex Bean）、峨眉红山茶（*C. omeiensis* Chang）、五柱滇山茶（*C. yunnanensis* Coh. St.）等种类的野生分布地。而且，张宏达教授于 1989 年和 1991 年先后发表了产自四川金沙江河谷地区的山茶属红山茶组新种就有 21 个。这些种类后来大部分被归并到云南山茶（*C. reticulata* Lindl.）和西南山茶，经细胞学研究，绝大部分是四倍体，少部分是六倍体，在盐边县还发现了二倍体的云南山茶。山茶在川渝地区栽培历史悠久，早在 1800 多年前的三国蜀汉时期山茶就已成为人工栽培的观赏花卉。明代（1621 年）王象晋的《群芳谱》中就有记载："山茶中宝珠为佳，蜀茶更胜"，突出了川渝地区的茶花在中国的地位。丰富的自然资源和悠久的栽培历史，造就了川渝地区丰富的山茶古树资源。

古树是不可再生和复制的稀缺资源，是祖先留下的宝贵财富，是中国培育山茶品种历史悠久的见证，许多古树本身就是传统名品，比如'杨妃茶''七心红''七心白''抓破脸'等。而且，这些古树也承载着悠久的历史和灿烂的文化，传递着世间的风云变幻和人间的沧桑，具有丰厚的文化内涵。研究和保护这些古树及其生态环境，在科研、生态、人文、旅游、经济、政治、历史以及丰富人们的文化生活和精神生活诸多方面都具有重要意义。

国际山茶协会（ICS）非常重视野生和栽培山茶古树和名木的保护，认为这些古树是千年人类文明中经济、园艺和文化成就的无价之宝和不可替代的象征。2018 年，国际山茶协会古树保育委员会正式成立。

2019年第二次委员会会议在广州召开，会议期间，发表了《山茶古树保护宣言》（简称《阿婆六宣言》），这是世界上首次提出的关于在全球范围内鼓励山茶古树名木保育的国际宣言。该宣言指定了山茶古树名木的认定标准，规范了古树的测量和记录，制定了国际山茶古树名木的评审制度和授牌程序，成为山茶古树保护的行动指南和纲领性文件。

《川山茶古树名录》是《阿婆六宣言》发表后，根据宣言精神出版的第一部山茶古树专著，而且本书的所有古树的名称都查对了世界山茶品种数据库，保证了名称的准确性和科学性。相信本书的出版对全面和系统地了解川渝地区的山茶古树资源有较大帮助，对我国茶花古树资源的有效保护和合理开发利用有十分重要的意义。希望以本书为基础，以地理信息系统为基础平台，建立川渝地区山茶古树的信息管理系统，对古树名木的生长环境、生长情况、保护现状等进行动态监测和跟踪管理，并定期向有关部门报告，充分发挥山茶古树在传承历史文化、弘扬生态文明中的独特作用，为推进绿色发展、建设美丽中国做出更大贡献。

国际山茶协会副主席
国际山茶属植物品种登录官
中国科学院昆明植物研究所研究员
2022年7月25日

Foreword II

The Sichuan-Chongqing region is an important distribution area of the genus *Camellia* in China, with abundant wild resources. According to incomplete statistics, there were more than 30 natural species of camellia distributed in this area. It is the wild distribution area of *Camellia pitardii, C. saluenensis, C. omeiensis, C. yunnanensis* and other species. In 1989 and 1991 only, professor Zhang Hongda (Hungta Chang) published 21 new species in Sect. Camellia of the genus *Camellia* in the Jinshajiang Valley of Sichuan Province, although most of them had been emerged into *C. reticulata* Lindl. and *C. pitardii.* We found that most of these new species were tetraploid, and a few were hexaploidy and diploid. The diploid plant of *Camellia reticulata* Lindl. was very precious which was also found in Yanbian County, Sichuan Province. Camellia has a long history of cultivation in Sichuan and Chongqing. Camellia has been cultivated as ornamental flowers more than 1,800 years ago during the Three Kingdoms period. In the book *Qunfangpu* (florilegium) written by Wang Xiangjin in 1621 (Ming Dynasty), had a record: "The 'Baozhu' is an excellent camellia, and the camellias in Shu (old place name, now Sichuan and Chongqing) are even better", which highlights the status of camellias in Sichuan and Chongqing in China. Rich natural resources and a long history of cultivation have created rich ancient camellia resources in Sichuan and Chongqing.

Ancient trees are scarce resources that could not be reproduced and copied, and are precious treasures left by our ancestors. They are witnesses to the long history of cultivating camellia cultivars in China. Many ancient trees are themselves traditional famous cultivars, such as 'Yangfei Cha' and 'Qixinhong', 'Qixinbai' and 'Zhuapolian'. Moreover, these ancient trees also carry a long history and splendid culture, convey the changes of the world and the vicissitudes of the world, and have rich cultural connotations. Studying and protecting these ancient trees and their ecological environment is of great significance in scientific research, ecology, humanities, tourism, economy, politics, history, and enriching people's cultural and spiritual life.

The International Camellia Society (ICS) has paid great attention to the conservation of both wild and cultivated ancient and historic camellia trees. These ancient trees are considered priceless and irreplaceable symbols of economic, horticultural and cultural achievements in the human civilization. The Committee of Historic for Camellia Conservation of the ICS was officially established in 2018. In 2019, its second committee meeting was held in Guangzhou of China. During the meeting, a Declaration for Historic Camellia Conservation (*Apoliu Declaration*) was announced. This is the world's first international declaration on encouraging the conservation of ancient and historic camellia trees on a global scale. The declaration specifies the standards for the identification of ancient and historic camellia trees, regulates the measurement and recording of ancient trees, and formulates the international evaluation system and licenses procedures for ancient and famous camellia trees, which has become an action guide and programmatic document for the protection of ancient camellia trees.

Records of Sichuan Ancient Camellia Trees is the first monograph of ancient camellia trees published in accordance with the spirit of The Declaration of Apoliu since it was published, and the names of all ancient trees in this book have been checked by the Database of International Camellia Register to ensure that the names of the ancient trees are accurate and scientific. I believed that the publication of this book is of great help to a comprehensive and systematic understanding of the ancient camellia resources in Sichuan and Chongqing, and is of great significance to the effective protection and rational development and utilization of ancient camellia resources in our country. I hoped that based on this book and the geographic information system as the basic platform, the information management system of ancient camellia trees in Sichuan and Chongqing could be established, to record the growth environment, growth situation and protection status of ancient and historic trees. This book will promote the unique role of ancient camellia trees in inheriting history and culture and promoting ecological civilization, and make greater contributions to promoting green development and building a beautiful China.

Prof. Wang Zhonglang
Vice president of International Camellia Society
International Camellia Registrar
Kunming Institute of Botany, CAS
25 July, 2022

前言

山茶花开于冬末春初，端庄圣洁，诗人郭沫若曾赞叹"茶花一树早桃红，百朵彤云啸傲中"。

2019年国际山茶协会（ICS）在《阿婆六宣言》中，"号召国际社会，特别是那些有山茶属植物生长的国家，包括其国家公园保护区、植物园、庄园、园艺和农业学会、遗产和文化协会等应对山茶科山茶属古树名木开展保护。"明确了原生山茶古树（就地或迁地古树）为树龄大于100年的植株，并提出了茶花古树的测量、记载及分级标准。本书中的川山茶古树调查记录及分级标准均采用该标准执行。

四川山茶简称川山茶，指主要分布于四川盆地及其周边地区的山茶花。四川山茶在川渝地区栽培历史悠久，其栽培历史可追溯至1800多年前的三国蜀汉时代。据上海著名园艺家黄德龄老先生的论断，"中国茶花起源于青藏高原，后分南北两支，南支发展为云南山茶，北支发展为四川山茶，嗣后四川山茶沿长江流域向东发展形成了华东山茶。这样就形成了四川山茶、云南山茶和华东山茶三大体系。"游慕贤老先生在《中国茶花古树觅踪十年》一书中指出，四川什邡、彭州、成都以及重庆等地有野生山茶和山茶古树资源，栽培型山茶花古树数量重庆居全国第二，其中重庆市南山植物园和万州西山公园保存的川山茶古树最多。

为进一步摸清川山茶古树在川渝地区的分布情况和保护现状，本书在我国第五次古树名木普查结果及前人调查的基础上，从2020年开始，调查了四川邛崃、峨眉山，重庆武陵山、四面山及中心城区等区域的川山茶古树。为准确测算树龄，本书请中国林科院亚热带林业研究所李纪元团队利用木质检测仪（IML-RESI PD 1000），采用微创打孔探测古树年轮，利用 PD-Tools Pro 软件分析计算树龄，同时结合史料记载，确定每株古树的树龄。

调查发现野生山茶花在四川省邛崃市天台山海拔 1000~1500 m 区域分布有上万株，峨眉山洪雅县周边海拔 800~1500 m 山上分布有近千株，在重庆市涪陵区双河乡和武隆区境内的后坪乡海拔 800~1800 m 分布有几千株，江津四面山海拔 1100~1600 m 分布有上千株，南川金佛山黄草坪海拔 1200~1800 m 分布有上千株，万州区恒河乡有少量分布，彭水县的龙射镇、新田镇、岩东乡等地周边海拔 800~1500 m 山上分布有几千株；这些野生型山茶古树种类主要有峨眉红山茶、西南红山茶、西南白山茶和多变性西南红山茶 4 种，调查发现多变性西南红山茶分布集中成片、规模最大的区域在彭水县新田镇，面积超过 6.7 hm^2；本书选取有代表性的野生型山茶花古树 59 株进行调查，其中树龄最大的是天台山的西南红山茶 279 年，植株最高大的是天台山的西南红山茶，株高 18 m，基围 188 cm；调查发现栽培型川山茶古树有 387 株，主要分布在重庆南山植物园（190 株）和万州西山公园（110 株），零星少量分布的有：四川成都、彭州和重庆周边地区，其品种主要有'紫金冠''醉杨妃''金顶大红'等 33 个，其中树龄最大的是重庆市南山植物园的'铁壳紫袍'，记载树龄有 400 多年；测量树龄最大的是万州西山公园的'紫金冠'，有 320 年。调查中发现四川省彭州市磁峰镇花塔村有两株千年古山茶'金盘荔枝'因自然原因已于 2014 年死亡，现枯桩仍在；重庆原江北区北庙子小学一株八百多年的川山茶古树'花五宝'于 1982 年因校园改建受损死亡，现枯桩存放在南山植物园山茶园贞寿亭内，以警示后人要热爱自然，保护古树。

本书承中国科学院昆明植物研究所研究员、国际山茶协会前主席管开云先生亲笔题词，中国林科院亚热带林业研究所研究员高继银先生，中国科学院昆明植物研究所研究员、国际山茶协会副主席王仲朗先生作序。在山茶古树调查过程中，得到了重庆市南山植物园管理处各级领导、中国林科院亚热带林业研究所李纪元研究员及其团队、昆明植物研究所杨世雄研究员、青城山—都江堰旅游景区管理局张瑞平、四川邛崃市盆景协会会长龚国文、邛崃市公园城市建设服务中心高级工程师顾顺祥、重庆缙云山自然保护区邓先宝研究员、西山公园魏映祥主任、西南大学教授李先源和杨晓红以及重庆师范大学教授唐安军、重庆市大足区城市管理局王宇、彭水县林业局高级工程师邓相舜、北京林业大学教授吴江梅等提供支持帮助，谨在此表示衷心感谢。

因新冠疫情防控影响，还有少部分调查没有实施，故本书未能全面反映川渝地区川山茶古树资源，敬请谅解。

编者
2022 年 5 月 20 日

Preface

A camellia blooms between the late winter and early spring and always displaces her dignified and holy manner. The point is supported by Chinese poet Guo Moruo in his poem "Earlier than peach blossom, the camellias stand proudly in their red-cloud-sea of flowers."

The International Camellia Society (ICS) published in 2019 its Declaration. It "commends to the global community, particularly those nations where plants of the genus Camellia are grown, including their national parks reserves and estates, botanical gardens, horticultural and agricultural societies, heritage and cultural associations, the conservation of ancient and historic trees of the family Theaceae, genus Camellia." (sic) It also explicitly defines that the protogenetic ancient camellia trees, or those in the original site or from ex-situ, should be over 100 years old. It also proposes the standards of measuring, recording and grading ancient camellia trees. In accordance with these standards, the investigating records and classifying methods of the Sichuan ancient camellia trees reported in this book have been implemented.

Sichuan Camellia refers to camellia mainly distributed in Sichuan Basin and its surrounding areas. It has a long cultivating history in the Sichuan-Chongqing region, and can be traced back to the period of Shu Kingdom (AD 221-263) of the Three Kingdoms (AD 220-280) more than 1,800 years ago. According to Mr. Huang Deling, a famous horticulturist in Shanghai, "Chinese camellia originated from the Qinghai-Tibet Plateau, and later it divided into north branch and south branch.

The south branch further developed into *Camellia reticulata* Lindl.; while the north branch evolved into Sichuan Camellia which extended continuously eastward along the Yangtze River to become East China Camellia. Thus three systems of Sichuan Camellia, *C. reticulata* Lindl. and East China Camellia have been formed." Mr. You Muxian points out in his book *Tracking down Chinese Ancient Camellia Trees for Ten Years* that there are wild camellias and ancient camellia resources in the cities of Shifang, Pengzhou and Chengdu of Sichuan Province (hereinafter Sichuan) and municipality Chongqing (hereinafter Chongqing). Chongqing ranks the second nationwide in the number of cultivated ancient camellia trees. The largest numbers of Sichuan ancient camellia trees are preserved in Chongqing Nanshan Botanical Garden and Wanzhou Xishan Park.

In order to further find out the distribution and protection status of the ancient camellia trees in Sichuan and Chongqing, based on the results of the fifth census of ancient and famous trees in China and previous investigations, we have surveyed the Sichuan ancient camellia trees in Qionglai City and Mount Emei of Sichuan Province and Wuling Mountain, Simian Mountain and the center district of Chongqing as well. To accurately measure the tree-age in the study, a wood detector (IML-RESI PD 1000) was applied with the technique of minimally invasive drilling to detect the growth rings of the ancient trees while PD-Tools Pro to analyze and calculate the tree age. The age of each ancient tree was finally determined after historical records were referred to. All the process was guided by our invited Li Jiyuan team of the Institute of Subtropical Forestry, CAF.

The investigation illustrated thousand wild camellias grow in Sichuan Tiantai Mountain at an elevation of 1000 - 1500 m, while almost one thousand wild camellias around Hongya County in Emeishan City at 800 - 1500 m a.s.l.. The distribution of other wild camellias include thousands in both Shuanghe Village of Fuling District and Houping Village of Wulong District in Chongqing City at 800 - 1800 m a.s.l., more than 1000 in Simian Mountain at 1100 - 1600 m a.s.l. , as well as that in Huangcaoping of Golden Buddha Mountain in Nanchuan District at 1200 ~ 1800 m a.s.l.. A small number develop in Henghe Village of Wanzhou District, with thousands scattering on the mountains at 800 ~ 1500 m a.s.l around Longshe Town, Xintian Town, Yandong Township and other places in Pengshui County. These wild types are mainly classified into the four species of *C. omeiensis* Chang, *C. pitardii* Coh.St., *C. pitardii* Coh. St. var. *alba* Chang and *C. pitardii* var. Compressa. The investigation found that the distribution of *C. pitardii* var. Compressa is in the form of clustering into patches and the largest area reaches more than 6.7 hm^2 in Xintian Town of Pengshui County. This book reports that 58 representative wild-type ancient camellia trees were selected for measurement and investigation, the oldest of which is the *C. pitardii* Coh.St. aged 227 in Tiantai Mountain. The tallest is also *C. pitardii* Coh.St. from Mount Tiantai, with a height of 18 m and a ground circumference of 60 cm. The research also manifests the finding of 367 cultivated-type Sichuan ancient camellia trees, including 190 in Chongqing Nanshan Botanical Garden, 110 in Wanzhou Xishan Park, and a small number in Chengdu and Pengzhou of Sichuan and the surrounding areas of Chongqing. They mainly contain 33 cultivars such as 'Zijinguan', 'Zuiyangfei' and 'Jinding Dahong'. Among them, the oldest tree is 'Tieke Zipao' in Chongqing Nanshan Botanical Garden, which is over 400 years old according to record, while the oldest of the measured tree is 'Zijinguan' in Wanzhou Xishan Park, aged 320. Two ancient camellias aged two thousand years old were found in the investigation from Huata Village of Cifeng County in Pengzhou City of Sichuan Although the two of the cultivars 'Jinpan Lizhi' had died in 2014 due to natural causes, the withered stumps still remain there. Another Sichuan ancient camellia tree 'Huawubao' aged over 800 died in 1982 of the campus reconstruction of Beimiaozi Primary School in the former Jiangbei District of Chongqing. The faded trunk is now stored in the Zhenshou Pavilion in the Camellia Garden of Nanshan Botanical Garden, which functions to warn future generations to love nature and protect ancient trees.

We would like to express our heartfelt thanks to writer of the inscription of this book, Mr. Guan Kaiyun, professor of Kunming Institute of Botany and former president of International Camellia Society and writers of the preface of this book, Mr. Gao Jiyin, professor of the Research Institute of Subtropical Forestry of CAF, and Mr. Wang Zhonglang, professor of the Kunming Institute of Botany of Chinese Academy of Sciences and vice president of International Camellia Society. We would also like to express our feeling of genuine gratitude towards all the following leaders and experts who provided great support and help in the process of our investigating ancient camellia trees: the leaders of Management Office of Chongqing Nanshan Botanical Garden, professor Li Jiyuan and his team of the Research Institute of Subtropical Forestry of CAF, Professor Yang Shixiong of Kunming Institute of Botany, Mr. Zhang Ruiping of Qingcheng Mountain-Dujiangyan Tourist Scenic Spot Administration, President Gong Guowen of Sichuan Qionglai Bonsai Association, Senior Engineer Gu Shunxiang of Park City Construction Service Center of Qionglai City, Professor Deng Xianbao of Jinyun Mountain Nature Reserve of Chongqing, Director Wei Yingxiang of Xishan Park, Professors Li Xianyuan and Yang Xiaohong of Chongqing Southwest University, Professor Tang Anjun of Chongqing Normal University, Mr. Wangyu of the Municipal Administration Bureau, Dazu District, Chongqing, Senior Engineer Deng Xiangshun of Pengshui County Forestry Bureau, and Professor Wu Jiangmei of Beijing Forestry University.

Due to the impact of COVID-19 pandemic prevention and control, a small number of investigations have not been conducted. Therefore please make allowance for the incomplete disclosure of the Sichuan ancient camellia resources in the areas of Sichuan and Chongqing.

Editors
20 May, 2022

四川省彭州市磁峰镇花塔村古山茶'金盘荔枝'原生长状况及现状图（因自然原因于2014年枯死）。
Growth status of ancient camellia tree years ago and recent in Huata Village, Cifeng Town, Pengzhou City, Sichuan Province (It died in 2014 due to natural causes).

重庆原江北区北庙子小学一株八百多年的川山茶古树'花五宝',因校园改建受损于 1982 年死亡,现枯桩存放于重庆南山植物园。
A more than 800-year-old sichuan ancient camellia tree 'Huawubao', which was formerly in the Beimiaozi Primary School of Jiangbei District in Chongqing, died in 1982 due to damage from campus reconstruction, and now the dead pile is stored in Chongqing Nanshan Botanical Garden.

目录 contents

序言 1

序言 2

前言

第一章　重庆地区山茶古树资源分布

重庆南山植物园	4
万州西山公园	106
万州区分水镇枣园	158
万州区恒合乡黄桐寨	160
西南大学	162
沙坪公园	180
重庆抗战遗址博物馆	196
鹅岭公园	206
北碚公园	212
花卉园	216
巴南区华林路	219
缙云寺	222
重庆市聚奎中学校	226
重庆金佛山国家级自然保护区	232

重庆武陵山国家森林公园	242
彭水苗族土家族自治县	256
四面山风景名胜区	262

第二章　四川地区山茶古树资源分布

都江堰离堆公园	272
雅安市及邛崃市	278
峨眉山景区万年寺	284
彭州市葛仙山风景区	286
邛崃市天台山风景区	288
峨眉山市	314

附表 324

参考文献 331

第一章
重庆地区山茶古树资源分布

Chapter 1
Distribution of Ancient Camellia Tree Resources in Chongqing region

　　重庆地区川山茶古树包括栽培型古树和野生型古树，本书详细记载了重庆地区369株栽培型山茶和27株野生型山茶。其中选取有代表性的单株227株古树进行详细介绍展示，其他的放于附表。因栽培型古树多数在公园、学校和私家栽植，移栽概率大，所以选取了部分树龄80至99年的大树作为后备古树资源收录于附表中。栽培型古树源于20世纪初兴建成的私家园林、公园和学校，如上世纪初伴随汪代玺、范崇实等一批归国留学生和实业家的私家花园修建，及公园和学校的兴建，栽植了大量的川山茶，并保存至今。川山茶古树集中分布于重庆南山植物园、万州西山公园、沙坪公园和西南大学内，其他地点有零散分布。川山茶古树的品种主要为'紫金冠''醉杨妃''七心红''重庆红''三学士'等32个，大部分珍贵的传统品种古树保存在南山植物园和西山公园内。现有川山茶古树除南山植物园和万州西山公园外，其他地点的生长状况整体较差，多表现为叶片黄绿、枝叶稀疏，甚至伴有明显枯死枝，需要加强管护。

　　野生型古树主要分布于重庆武陵山、金佛山、四面山、阴条岭、茂云山和七曜山等大山之中，海拔800~1800 m之间。以西南红山茶、西南白山茶和多变性西南红山茶为主，生长健壮。

The Sichuan ancient camellia trees in Chongqing include cultivated type and wild type. This book records 369 cultivated trees and 27 wild trees in Chongqing area in detail, among which, 227 typical individual trees are selected for detailed introduction, while the remaining are listed in the supplemental table. Most of the cultivated type are preserved in parks, schools, and private gardens, which have high possibilities of being transplanted. Thus, this book selected and recorded some additional old trees with an age of 80 to 99 years as ancient tree's reserve resources in the supplemental table. The cultivated type originated from the construction of private gardens, parks, and schools in early twenty century and was preserved up to now. For instance, the construction of private gardens for returning students after studying abroad and industrialists such as Wang Daixi and Fan Chongshi, as well as the construction of parks and schools early twenty century, a large amount of Sichuan camellia trees which were planted and preserved up to now. The distribution of Sichuan ancient camellia trees are concentrated in Chongqing Nanshan Botanical Garden, Wanzhou Xishan Park, Shaping Park and Southwest University, and are also scattered in other locations. There are 32 varieties of Sichuan ancient camellia trees, including 'Zijinguan' 'Zuiyangfei', 'Qixinhong' 'Chongqinghong', and 'Sanxueshi'. Most of the precious traditional varieties of ancient trees are only preserved at Nanshan Botanical Garden and Xishan Park. Except for those grown at Nanshan Botanical Garden and Wanzhou Xishan Park, the remaining Sichuan ancient camellia trees are overall in weakly grown, mainly shown as sparse branches and foliage, unhealthy yellow-green foliage, and even with obvious dead branches. Thus, management and caretaking of those trees need to be improved.

Wild-type trees are mainly distributed in Wuling Mountain, Golden Buddha Mountain, Simian Mountain, Yintiaoling, Maoyun Mountain and Qiyao Mountain in Chongqing with an altitude of 800 - 1800 m, majority of which are *C. pitaradii* coh. St., *C. pitardii* var. *alba* Chang and *C. pitardii* var. Compressa and grow robustly.

重庆
南山植物园

*INTRODUCTION TO CHONGQING
NANSHAN BOTANICAL GARDEN*

重庆市南山植物园建于1959年，是在原民国时期留法医生汪代玺和民族实业家范崇实的私家花园基础上改建而成，园内川山茶古树主要来源于原私家花园、江北龙溪苗圃和北碚静观镇等地。山茶专类园于2004年3月建成并对外开放，因山茶花为重庆市花，故该园又称市花园。重庆市南山植物园作为川山茶种质资源的重要保存地，现有川山茶古树190棵，2012年被国际山茶花协会评为"国际杰出茶花园"，2016年入选首批国家山茶种质资源库，2022年入选重庆市首批历史名园。由于部分古茶因茶花园改建，树势有所衰退，自2005年起公园对古茶树开展了增施生物有机肥，提高土壤有益菌含量、深翻、打透气孔等土壤改良措施；适时修剪，喷施酸性叶面肥，根施微量元素等复壮处理以增强根系活力。

　　Chongqing Nanshan Botanical Garden was built in 1959, which was originally the private gardens of Wang Daixi, a doctor study in France in the Republic of China, and of Fan Chongshi, a national industrialist. The Sichuan ancient camellia trees in the park mainly included those preserved from the original private gardens, as well as those transplanted from Jiangbei Longxi Nursery and Beibei Jingguan Town. The Camellia Garden was built and opened to the public in March 2004. Because camellia is the representative flower of Chongqing, the garden is also called the City Garden. Chongqing Nanshan Botanical Garden, as an important preservation place for Sichuan Camellia germplasm resources, currently has 190 ancient camellia trees. In 2012, it was awarded as "International Camellia Garden of Excellence" by International Camellia Society. In 2016, it was selected into the first batch of National Camellia Germplasm Resource Center. In 2022, the park was selected into the first batch of famous historic parks in Chongqing. Due to the reconstruction of the Camellia Garden, some ancient camellia trees were affected with decreased growth potential. To improve growth condition of camellia trees and restore soil quality, the park management office had started to fertilize ancient camellia trees with bio-organic fertilizers to increase beneficial bacteria content in the soil, as well as digging deeply and drilling air holes around camellia trees since 2005. By the same time, the management office also takes good care of camellia trees by properly prune the side branches, periodically spray acidic foliar fertilizer, as well as restore root vitality by supplementing the root with elements such as iron, magnesium and others upon needs.

古1：花洋红
C. japonica 'Hua Yanghong'

位于重庆市南山植物园古茶苑，N29°33′26″，E106°37′19″，海拔 540.6 m。株高 5.3 m，基围 58.0 cm，两分枝 53.5/37.5 cm，冠幅 5.0 m，树龄 164 年，花期 1–4 月，是川山茶传统品种，长势良好。管护单位重庆市南山植物园管理处。

Located in the Gucha Yard, Chongqing Nanshan Botanical Garden, with latitude 29°33′26″ N, longitude 106°37′19″ E, elevation 540.6 m, tree H 5.3 m, CG 58.0 cm, two branches (53.5 cm /37.5 cm), CD 5.0 m, 164 y, Fl. Jan. to Apr. It is a traditional cultivar of Sichuan Camellia and grows well. Managed and maintained by Chongqing Nanshan Botanical Garden Administrative Office.

古2：七心红
C. japonica 'Qixinhong'

位于重庆市南山植物园古茶苑，N29°33′26″，E106°37′19″，海拔540.8 m。株高4.0 m，基围75.0 cm，五分枝，最大三分枝34.0/35.0/32.0 cm，冠幅6.0 m，树龄172年，花期1–4月，长势良好。管护单位重庆市南山植物园管理处。

Located in the Gucha Yard, Chongqing Nanshan Botanical Garden, with latitude 29°33′26″ N, longitude 106°37′19″ E, elevation 540.8 m, tree H 4.0 m, CG 75.0 cm, five branches (the biggest three 34.0 cm/35.0 cm/32.0 cm respectively), CD 6.0 m, 172 y, Fl. Jan. to Apr. Grows well. Managed and maintained by Chongqing Nanshan Botanical Garden Administrative Office.

古3：九心十八瓣
C. japonica 'Jiuxin Shibaban'

位于重庆市南山植物园古茶苑，N29°33′26″，E106°37′19″，海拔542.7 m。株高6.0 m，基围60.0 cm，冠幅3.5 m，树龄138年，花期2–4月，是川山茶传统名贵品种，长势良好。管护单位重庆市南山植物园管理处。

Located in the Gucha Yard, Chongqing Nanshan Botanical Garden, with latitude 29°33′26″ N, longitude 106°37′19″ E, elevation 542.7 m, tree H 6.0 m, CG 60.0 cm, CD 3.5 m, 138 y, Fl. Feb. to Apr. It is a traditional precious cultivar of Sichuan Camellia and grows well. Managed and maintained by Chongqing Nanshan Botanical Garden Administrative Office.

古 4: 白宝塔
C. japonica 'Baibaota'

位于重庆市南山植物园古茶苑，N29°33′26″，E106°37′19″，海拔 540.6 m。株高 4.5 m，基围 70.0 cm，四分枝，最大三分枝 36.0/37.0/41.0 cm，冠幅 6.0 m，树龄 139 年，花期 2–4 月，长势良好。管护单位重庆市南山植物园管理处。

Located in the Gucha Yard, Chongqing Nanshan Botanical Garden, with latitude 29°33′26″ N, longitude 106°37′19″ E, elevation 540.6 m, tree H 4.5 m, CG 70.0 cm, four branches (the biggest three 36.0 cm/37.0 cm/41.0 cm respectively), CD 6.0 m, 139 y, Fl. Feb. to Apr. Grows well. Managed and maintained by Chongqing Nanshan Botanical Garden Administrative Office.

古5：七心红
C. japonica 'Qixinhong'

位于重庆市南山植物园古茶苑，N29°33′26″，E106°37′19″，海拔 540.9 m。株高 3.6 m，基围 75.0 cm，四分枝，最大两分枝 44.0/44.0 cm，冠幅 6.0 m，树龄 172 年，花期 1–4 月，长势良好。管护单位重庆市南山植物园管理处。

Located in the Gucha Yard, Chongqing Nanshan Botanical Garden, with latitude 29°33′26″ N, longitude 106°37′19″ E, elevation 540.9 m, tree H 3.6 m, CG 75.0 cm, four branches (the biggest two 44.0 cm/44.0 cm respectively), CD 6.0 m, 172 y, Fl. Jan. to Apr. Grows well. Managed and maintained by Chongqing Nanshan Botanical Garden Administrative Office.

古6: 川牡丹茶
C. japonica 'Chuan Mudan Cha'

位于重庆市南山植物园古茶苑，N29°33′26″，E106°37′19″，海拔 540.9 m。株高 4.7 m，基围 66.0 cm，两分枝 60.0/25.0 cm，冠幅 4.3 m，树龄 153 年，是川山茶传统名贵品种，也是花期最早、花径最大的品种。花期 10 月至翌年 4 月，长势良好。管护单位重庆市南山植物园管理处。

Located in the Gucha Yard, Chongqing Nanshan Botanical Garden, with latitude 29°33′26″ N, longitude 106°37′19″ E, elevation 540.9 m, tree H 4.7 m, CG 66.0 cm, two branches (60.0 cm /25.0 cm), CD 4.3 m, 153 y. It is a traditional precious cultivar of Sichuan Camellia with the earliest florescence and the largest flower diameter. Fl. Oct. to Apr. Grows well. Managed and maintained by Chongqing Nanshan Botanical Garden Administrative Office.

古 7：三学士
C. japonica 'Sanxueshi'

位于重庆市南山植物园古茶苑，N29°33′26″，E106°37′19″，海拔 540.6 m。株高 4.0 m，基围 50.0 cm，两分枝 36.0/32.0 cm，冠幅 4.0 m，树龄 138 年，是川山茶传统名贵、晚花品种，花期 3–5 月初，长势良好。管护单位重庆市南山植物园管理处。

Located in the Gucha Yard, Chongqing Nanshan Botanical Garden, with latitude 29°33′26″ N, longitude 106°37′19″ E, elevation 540.6 m, tree H 4.0 m, CG 50.0 cm, two branches (36.0 cm/32.0 cm), CD 4.0 m, 138 y. It is a traditional precious cultivar of Sichuan camellia with relatively late blooming period. Fl. Mar. to May. Grows well. Managed and maintained by Chongqing Nanshan Botanical Garden Administrative Office.

古8：白宝塔
C. japonica 'Baibaota'

位于重庆市南山植物园古茶苑，N29°33′26″，E106°37′19″，海拔 540.9 m。株高 4.8 m，基围 47.0 cm，两分枝 36.0/35.0 cm，冠幅 4.5 m，树龄 139 年，花期 2–4 月，长势良好。管护单位重庆市南山植物园管理处。

Located in the Gucha Yard, Chongqing Nanshan Botanical Garden, with latitude 29°33′26″ N, longitude 106°37′19″ E, elevation 540.9 m, tree H 4.8 m, CG 47.0 cm, two branches (36.0 cm/35.0 cm), CD 4.5 m, 139 y. Fl. Feb. to Apr. Grows well. Managed and maintained by Chongqing Nanshan Botanical Garden Administrative Office.

古9：白洋片
C. japonica 'Baiyangpian'

位于重庆市南山植物园古茶苑，N29°33′27″，E106°37′19″，海拔 532.2 m。株高 5.3 m，基围 55.0 cm，两分枝 32.0/36.0 cm，冠幅 4.5 m，树龄 112 年，花期 2–4 月，长势良好。管护单位重庆市南山植物园管理处。

Located in the Gucha Yard, Chongqing Nanshan Botanical Garden, with latitude 29°33′27″ N, longitude 106°37′19″ E, elevation 532.2 m, tree H 5.3 m, CG 55.0 cm, two branches (32.0 cm/36.0 cm), CD 4.5 m, 112 y. Fl. Feb. to Apr. Grows well. Managed and maintained by Chongqing Nanshan Botanical Garden Administrative Office.

古10：七心红
C. japonica 'Qixinhong'

位于重庆市南山植物园古茶苑，N29°33′26″，E106°37′19″，海拔540.9 m。株高3.6 m，基围51.0 cm，三分枝30.0/27.0/28.0 cm，冠幅5.0 m，树龄113年，花期1–4月，长势良好。管护单位重庆市南山植物园管理处。

Located in the Gucha Yard, Chongqing Nanshan Botanical Garden, with latitude 29°33′26″ N, longitude 106°37′19″ E, elevation 540.9 m, tree H 3.6 m, CG 51.0 cm, three branches (30.0 cm/27.0 cm/28.0 cm), CD 5.0 m, 113 y, Fl. Jan. to Apr. Grows well. Managed and maintained by Chongqing Nanshan Botanical Garden Administrative Office.

古11：七心红
C. japonica 'Qixinhong'

位于重庆市南山植物园古茶苑，N29°33′26″，E106°37′19″，海拔 541.1 m。株高 3.6 m，基围 55.0 cm，三分枝 30.0/27.0/30.0 cm，冠幅 3.6 m，树龄 113 年，花期 1–4 月，长势良好。管护单位重庆市南山植物园管理处。

Located in the Gucha Yard, Chongqing Nanshan Botanical Garden, with latitude 29°33′26″ N, longitude 106°37′19″ E, elevation 541.1 m, tree H 3.6 m, CG 55.0 cm, three branches (30.0 cm/27.0 cm/30.0 cm), CD 3.6 m, 113 y, Fl. Jan. to Apr. Grows well. Managed and maintained by Chongqing Nanshan Botanical Garden Administrative Office.

古 12：川牡丹茶
C. japonica 'Chuan Mudan Cha'

位于重庆市南山植物园古茶苑，N29°33′26″，E106°37′19″，海拔 541.0 m。株高 3.8 m，基围 55.0 cm，冠幅 3.0 m，树龄 139 年，是川山茶传统名贵品种，也是花期最早、花径最大的川茶品种，花期 10 月至翌年 4 月，长势良好。管护单位重庆市南山植物园管理处。

Located in the Gucha Yard, Chongqing Nanshan Botanical Garden, with latitude 29°33′26″ N, longitude 106°37′19″ E, elevation 541.0 m, tree H 3.8 m, CG 55.0 cm, CD 3.0 m, 139 y. It is a traditional precious cultivar of Sichuan Camellia with the earliest florescence and the largest flower diameter. Fl. Oct. to Apr. Grows well. Managed and maintained by Chongqing Nanshan Botanical Garden Administrative Office.

古 13：石榴茶
C. japonica 'Shiliu Cha'

位于重庆市南山植物园古茶苑，N29°33′26″，E106°37′18″，海拔 541.1 m。株高 6.0 m，基围 81.0 cm，四分枝，最大三个分枝 44.0/47.0/35.0 cm，冠幅 6.0 m，树龄 193 年，是川山茶传统名贵品种，花期 1–4 月，长势良好。管护单位重庆市南山植物园管理处。

Located in the Gucha Yard, Chongqing Nanshan Botanical Garden, with latitude 29°33′26″ N, longitude 106°37′18″ E, elevation 541.1 m, tree H 6.0 m, CG 81.0 cm, four branches (the biggest three 44.0 cm/47.0 cm/35.0 cm respectively), CD 6.0 m, 193 y. It is a traditional precious cultivar of Sichuan Camellia. Fl. Jan. to Apr. Grows well. Managed and maintained by Chongqing Nanshan Botanical Garden Administrative Office.

古 14：绒团茶
C. japonica 'Rongtuancha'

位于重庆市南山植物园古茶苑，N29°33′26″，E106°37′18″，海拔 541.0 m。株高 5.8 m，基围 74.0 cm，两分枝 50.0/55.0 cm，冠幅 6.8 m，树龄 107 年，花期 2–4 月，长势良好。管护单位重庆市南山植物园管理处。

Located in the Gucha Yard, Chongqing Nanshan Botanical Garden, with latitude 29°33′26″ N, longitude 106°37′18″ E, elevation 541.0 m, tree H 5.8 m, CG 74.0 cm, two branches (50.0 cm/55.0 cm), CD 6.8 m, 107 y, Fl. Feb. to Apr. Grows well. Managed and maintained by Chongqing Nanshan Botanical Garden Administrative Office.

古 15：抓破脸
C. japonica 'Zhuapolian'

位于重庆市南山植物园古茶苑，N29°33′26″，E106°37′18″，海拔 541.0 m。株高 3.5 m，基围 56.0 cm，两分枝 35.0/45.0 cm，冠幅 4.5 m，树龄 121 年，花期 2-4 月，长势良好。管护单位重庆市南山植物园管理处。

Located in the Gucha Yard, Chongqing Nanshan Botanical Garden, with latitude 29°33′26″ N, longitude 106°37′18″ E, elevation 541.0 m, tree H 3.5 m, CG 56.0 cm, two branches (35.0 cm/45.0 cm), CD 4.5 m, 121 y, Fl. Feb. to Apr. Grows well. Managed and maintained by Chongqing Nanshan Botanical Garden Administrative Office.

古16：七心红
C. japonica 'Qixinhong'

位于重庆市南山植物园古茶苑，N29°33′26″，E106°37′18″，海拔541.0 m。株高3.0 m，基围90.5 cm，基部膨大，冠幅5.0 m，树龄172年，花期1-4月，长势良好。管护单位重庆市南山植物园管理处。

Located in the Gucha Yard, Chongqing Nanshan Botanical Garden, with latitude 29°33′26″ N, longitude 106°37′18″ E, elevation 541.0 m, tree H 3.0 m, CG 90.5 cm, swollen base, CD 5.0 m, 172 y, Fl. Jan. to Apr. Grows well. Managed and maintained by Chongqing Nanshan Botanical Garden Administrative Office.

古17：抓破脸
C. japonica 'Zhuapolian'

位于重庆市南山植物园古茶苑，N29°33′26″，E106°37′18″，海拔541.0 m。株高3.0 m，基围52.0 cm，三分枝32.5/25.0/22.0 cm，冠幅4.6 m，树龄121年，花期2–4月，长势良好。管护单位重庆市南山植物园管理处。

Located in the Gucha Yard, Chongqing Nanshan Botanical Garden, with latitude 29°33′26″ N, longitude 106°37′18″ E, elevation 541.0 m, tree H 3.0 m, CG 52.0 cm, three branches (32.5 cm/25.0 cm/22.0 cm), CD 4.6 m, 121 y, Fl. Feb. to Apr. Grows well. Managed and maintained by Chongqing Nanshan Botanical Garden Administrative Office.

古18：小红莲
C. japonica 'Xiaohonglian'

位于重庆市南山植物园图书馆往八德亭路边，N29°33′28″，E106°37′22″，海拔535.8 m。株高5.0 m，基围69.5 cm，三分枝41.0/31.0/28.5 cm，冠幅5.0 m，树龄138年，花期3–5月初，长势良好。管护单位重庆市南山植物园管理处。

Located at the roadside from library to Bade Pavilion, Chongqing Nanshan Botanical Garden, with latitude 29°33′28″ N, longitude 106°37′22″ E, elevation 535.8 m, tree H 5.0 m, CG 69.5 cm, three branches (41.0 cm/31.0 cm/28.5 cm), CD 5.0 m, 138 y, Fl. Mar. to May. Grows well. Managed and maintained by Chongqing Nanshan Botanical Garden Administrative Office.

古19：金顶大红
C. japonica 'Jinding Dahong'

位于重庆市南山植物园图书馆往八德亭路边，N29°33′28″，E106°37′23″，海拔534.8 m。株高4.8 m，基围65.5 cm，两分枝44.0/41.0 cm，冠幅5.3 m，树龄139年，花期1–4月，长势良好。管护单位重庆市南山植物园管理处。

Located at the roadside from library to Bade Pavilion, Chongqing Nanshan Botanical Garden, with latitude 29°33′28″ N, longitude 106°37′23″ E, elevation 534.8 m, tree H 4.8 m, CG 65.5 cm, two branches (44.0 cm/41.0 cm), CD 5.3 m, 139 y, Fl. Jan. to Apr. Grows well. Managed and maintained by Chongqing Nanshan Botanical Garden Administrative Office.

古 20: 胭脂鳞
C. japonica 'Yanzhilin'

位于重庆市南山植物园图书馆往八德亭路边，N29°33′28″，E106°37′23″，海拔 533.8 m。株高 4.3 m，基围 54.0 cm，冠幅 4.5 m，树龄 164 年，花期 1–4 月，长势良好。管护单位重庆市南山植物园管理处。

Located at the roadside from library to Bade Pavilion, Chongqing Nanshan Botanical Garden, with latitude 29°33′28″ N, longitude 106°37′23″ E, elevation 533.8 m, tree H 4.3 m, CG 54.0 cm, CD 4.5 m, 164 y, Fl. Jan. to Apr. Grows well. Managed and maintained by Chongqing Nanshan Botanical Garden Administrative Office.

古21：胭脂鳞
C. japonica 'Yanzhilin'

位于重庆市南山植物园紫金冠大道，N29°33′28″、E106°37′23″，海拔533.4 m。株高4.5 m，基围58.0 cm，冠幅3.5 m，树龄164年，花期1–4月，长势良好。管护单位重庆市南山植物园管理处。

Located at the side of Zijinguan Road, Chongqing Nanshan Botanical Garden, with latitude 29°33′28″ N, longitude 106°37′23″ E, elevation 533.4 m, tree H 4.5 m, CG 58.0 cm, CD 3.5 m, 164 y, Fl. Jan. to Apr. Grows well. Managed and maintained by Chongqing Nanshan Botanical Garden Administrative Office.

古22：白洋片
C. japonica 'Baiyangpian'

位于重庆市南山植物园紫金冠大道，N29°33′28″，E106°37′24″，海拔528.3 m。株高4.8 m，基围71.5 cm，三分枝36.0/27.5/42.5 cm，冠幅4.6 m，树龄112年，花期2–4月，长势良好。管护单位重庆市南山植物园管理处。

Located at the side of Zijinguan Road, Chongqing Nanshan Botanical Garden, with latitude 29°33′28″ N, longitude 106°37′24″ E, elevation 528.3 m, tree H 4.8 m, CG 71.5 cm, three branches (36.0 cm/27.5 cm/42.5 cm), CD 4.6 m, 112 y, Fl. Feb. to Apr. Grows well. Managed and maintained by Chongqing Nanshan Botanical Garden Administrative Office.

古 23：九心十八瓣
C. japonica 'Jiuxin Shibaban'

位于重庆市南山植物园紫金冠大道底端，N29°33′28″，E106°37′24″，海拔527.0 m。株高5.3 m，基围51.5 cm，冠幅4.0 m，树龄139年，花期2–4月，长势良好。管护单位重庆市南山植物园管理处。

Located at the end of Zijinguan Road, Chongqing Nanshan Botanical Garden, with latitude 29°33′28″ N, longitude 106°37′24″ E, elevation 527.0 m, tree H 5.3 m, CG 51.5 cm, CD 4.0 m, 139 y, Fl. Feb. to Apr. Grows well. Managed and maintained by Chongqing Nanshan Botanical Garden Administrative Office.

古 24：绒团茶
C. japonica 'Rongtuancha'

位于重庆市南山植物园古茶苑靠宏建山庄围墙处，N29°33′26″，E106°37′19″，海拔 540.9 m。株高 5.5 m，基围 74.5 cm，三分枝，最大两分枝 43.0/48.5 cm，冠幅 5.5 m，树龄 107 年。花期 2–4 月，长势良好。管护单位重庆市南山植物园管理处。

Located in the Gucha Yard near the wall of Hongjian Villa, Chongqing Nanshan Botanical Garden, with latitude 29°33′26″ N, longitude 106°37′19″ E, elevation 540.9 m, tree H 5.5 m, CG 74.5 cm, three branches (the biggest two 43.0 cm/48.5 cm respectively), CD 5.5 m, 107 y, Fl. Feb. to Apr. Grows well. Managed and maintained by Chongqing Nanshan Botanical Garden Administrative Office.

古 25：七心红
C. japonica 'Qixinhong'

位于重庆市南山植物园古茶苑，N29°33′26″，E106°37′19″，海拔 542.2 m。株高 3.0 m，基围 63.0 cm，六分枝，最大两分枝 33.0/29.0 cm，冠幅 4.3 m，树龄 172 年。花期 1–4 月，长势良好。管护单位重庆市南山植物园管理处。

Located in the Gucha Yard, Chongqing Nanshan Botanical Garden, with latitude 29°33′26″ N, longitude 106°37′19″ E, elevation 542.2 m, tree H 3.0 m, CG 63.0 cm, six branches (the biggest two 33.0 cm /29.0 cm respectively), CD 4.3 m, 172 y, Fl. Jan. to Apr. Grows well. Managed and maintained by Chongqing Nanshan Botanical Garden Administrative Office.

古26：川牡丹茶
C. japonica 'Chuan Mudan Cha'

位于重庆市南山植物园古茶苑，N29°33′26″，E106°37′19″，海拔540.9 m。株高4.3 m，基围61.5 cm，两分枝36.0/53.5 cm，冠幅5.3 m，树龄153年，花期10月至翌年4月，长势良好。管护单位重庆市南山植物园管理处。

Located in the Gucha Yard, Chongqing Nanshan Botanical Garden, with latitude 29°33′26″ N, longitude 106°37′19″ E, elevation 540.9 m, tree H 4.3 m, CG 61.5 cm, two branches (36.0 cm /53.5 cm), CD 5.3 m, 153 y, Fl. Oct. to Apr. Grows well. Managed and maintained by Chongqing Nanshan Botanical Garden Administrative Office.

古27: 醉杨妃
C. japonica 'Zuiyangfei'

位于重庆市南山植物园古茶苑，N29°33′26″，E106°37′19″，海拔 540.4 m。株高 4.3 m，基围 51.0 cm，三分枝 37.0/30.0/27.0 cm，冠幅 4.8 m，树龄 112 年，花期 2–4 月，长势良好。管护单位重庆市南山植物园管理处。

Located in the Gucha Yard, Chongqing Nanshan Botanical Garden, with latitude 29°33′26″ N, longitude 106°37′19″ E, elevation 540.4 m, tree H 4.3 m, CG 51.0 cm, three branches (37.0 cm/30.0 cm/27.0 cm), CD 4.8 m, 112 y, Fl. Feb. to Apr. Grows well. Managed and maintained by Chongqing Nanshan Botanical Garden Administrative Office.

古 28：川牡丹茶
C. japonica 'Chuan Mudan Cha'

位于重庆市南山植物园古茶苑圆门处，N29°33′26″，E106°37′19″，海拔540.8 m。株高3.5 m，基围63.0 cm，四分枝25.5/33.5/24.0/23.5 cm，冠幅3.8 m，树龄153年，花期10月至翌年4月，长势良好。管护单位重庆市南山植物园管理处。

Located at the round gate of fhe Gucha Yard, Chongqing Nanshan Botanical Garden, with latitude 29°33′26″ N, longitude 106°37′19″ E, elevation 540.8 m, tree H 3.5 m, CG 63.0 cm, four branches (25.5 cm/33.5 cm/24.0 cm/23.5 cm), CD 3.8 m, 153 y, Fl. Oct. to Apr. Grows well. Managed and maintained by Chongqing Nanshan Botanical Garden Administrative Office.

古 29：铁壳紫袍
C. japonica 'Tieke Zipao'

　　位于重庆市南山植物园古茶苑灿若云霞处，N29°33′26″，E106°37′19″，海拔 539.0 m。株高 2.2 m，基围 43.0 cm，两分枝，已枯死一枝，冠幅 2.0 m，树龄 164 年，花期 2–4 月。树干基部空腐严重，长势一般。管护单位重庆市南山植物园管理处。

　　Located at the Canruoyunxia of the Gucha Yard, Chongqing Nanshan Botanical Garden, with latitude 29°33′26″ N, longitude 106°37′19″ E, elevation 539.0 m, tree H 2.2 m, CG 43.0 cm, two branches but one is dead, CD 2.0 m, 164 y, Fl. Feb. to Apr. Sever hollow rot at the base of the trunk. Managed and maintained by Chongqing Nanshan Botanical Garden Administrative Office.

古30：铁壳紫袍
C. japonica 'Tieke Zipao'

位于重庆市南山植物园古茶苑灿若云霞处，N29°33′26″，E106°37′19″，海拔539.0 m。株高3.0 m，基围90.0 cm，四分枝，最大三个分枝32.0/23.5/28.5 cm，冠幅3.5 m，花期2-4月。据史料记载，树龄400多年，主干早已空腐，剩一张树皮支撑，在植物园重点管护下，长势良好。管护单位重庆市南山植物园管理处。

Located at the Canruoyunxia of the Gucha Yard, Chongqing Nanshan Botanical Garden, with latitude 29°33′26″ N, longitude 106°37′19″ E, elevation 539.0 m, tree H 3.0 m, CG 90.0 cm, four branches (the biggest three 32.0 cm/23.5 cm/28.5 cm respectively), CD 3.5 m, Fl. Feb. to Apr. According to historical records, it is more than 400 years old. The trunk is rotted and empty, only with the bark left for survival. It is growing well after careful and focused maintainess of Nanshan Botanical Garden. Managed and maintained by Chongqing Nanshan Botanical Garden Administrative Office.

古 31：黑艳红
C. japonica 'Heiyanhong'

位于重庆市南山植物园古茶苑灿若云霞处，N29°33′26″，E106°37′20″，海拔 537.6 m。株高 4.0 m，基围 54.0 cm，三分枝，最大两分枝 33.0/34.0 cm，冠幅 5.0 m，树龄 164 年，花黑红色，花期 2-4 月，长势良好。管护单位重庆市南山植物园管理处。

Located at the Canruoyunxia of the Gucha Yard, Chongqing Nanshan Botanical Garden, with latitude 29°33′26″ N, longitude 106°37′20″ E, elevation 537.6 m, tree H 4.0 m, CG 54.0 cm, three branches (the biggest two 33.0 cm/34.0 cm respectively), CD 5.0 m, 164 y, flowers are black and red, Fl. Feb. to Apr. Grows well. Managed and maintained by Chongqing Nanshan Botanical Garden Administrative Office.

古32: 三学士
C. japonica 'Sanxueshi'

位于重庆市南山植物园古茶苑灿若云霞处，N29°33′26″，E106°37′19″，海拔 538.8 m。株高 5.0 m，基围 76.0 cm，三分枝 61.0/30.0/32.0 cm，冠幅 4.7 m，树龄 190 年，花期 3–5 月，长势良好。管护单位重庆市南山植物园管理处。

Located at the Canruoyunxia of the Gucha Yard, Chongqing Nanshan Botanical Garden, with latitude 29°33′26″ N, longitude 106°37′19″ E, elevation 538.8 m, tree H 5.0 m, CG 76.0 cm, three branches (61.0 cm/30.0 cm/32.0 cm), CD 4.7 m, 190 y, Fl. Mar. to May. Grows well. Managed and maintained by Chongqing Nanshan Botanical Garden Administrative Office.

古 33：紫金冠
C. japonica 'Zijinguan'

位于重庆市南山植物园古茶苑灿若云霞处，N29°33′25″，E106°37′19″，海拔538.0 m。株高4.5 m，基围55.0 cm，两分枝36.5/36.0 cm，冠幅4.4 m，树龄169年，花期1–4月，长势一般。管护单位重庆市南山植物园管理处。

Located at the Canruoyunxia of the Gucha Yard, Chongqing Nanshan Botanical Garden, with latitude 29°33′25″ N, longitude 106°37′19″ E, elevation 538.0 m, tree H 4.5 m, CG 55.0 cm, two branches (36.5 cm/36.0 cm), CD 4.4 m, 169 y, Fl. Jan. to Apr. Managed and maintained by Chongqing Nanshan Botanical Garden Administrative Office.

古34：紫金冠
C. japonica 'Zijinguan'

位于重庆市南山植物园古茶苑灿若云霞与宏建山庄之间，N29°33′25″，E106°37′19″，海拔538.0 m。株高5.5 m，基围68.0 cm，冠幅5.0 m，树龄169年，花期1–4月，长势良好。管护单位重庆市南山植物园管理处。

Located between the Canruoyunxia and the Hongjian Villa, Chongqing Nanshan Botanical Garden, with latitude 29°33′25″ N, longitude 106°37′19″ E, elevation 538.0 m, tree H 5.5 m, CG 68.0 cm, CD 5.0 m, 169 y, Fl. Jan. to Apr. Grows well. Managed and maintained by Chongqing Nanshan Botanical Garden Administrative Office.

古35：花洋红
C. japonica 'Hua Yanghong'

位于重庆市南山植物园古茶苑灿若云霞处，N29°33′25″，E 106°37′19″，海拔 537.2 m。株高 4.2 m，基围 49.5 cm，两分枝 26.0/30.0 cm，冠幅 3.5 m，树龄 107 年，花期 2–4 月，长势良好。管护单位重庆市南山植物园管理处。

Located at the Canruoyunxia of the Gucha Yard, Chongqing Nanshan Botanical Garden, with latitude 29°33′25″ N, longitude 106°37′19″ E, elevation 537.2 m, tree H 4.2 m, CG 49.5 cm, two branches (26.0 cm/30.0 cm), CD 3.5 m, 107 y, Fl. Feb. to Apr. Grows well. Managed and maintained by Chongqing Nanshan Botanical Garden Administrative Office.

古 36：金盘荔枝
C. japonica 'Jinpan Lizhi'

位于重庆市南山植物园古茶遗韵处，N29°33′25″，E106°37′19″，海拔 534.5 m。株高 4.5 m，基围 54.0 cm，冠幅 4.2 m，树龄 112 年，花期 2–4 月，长势良好。管护单位重庆市南山植物园管理处。

Located at the Guchayiyun, Chongqing Nanshan Botanical Garden, with latitude 29°33′25″ N, longitude 106°37′19″ E, elevation 534.5 m, tree H 4.5 m, CG 54.0 cm, CD 4.2 m, 112 y, Fl. Feb. to Apr. Grows well. Managed and maintained by Chongqing Nanshan Botanical Garden Administrative Office.

古37：九心十八瓣
C. japonica 'Jiuxin Shibaban'

位于重庆市南山植物园古茶遗韵处，N29°33'25"，E106°37'19"，海拔535.7 m。株高4.8 m，两分枝43.0/27.3 cm，冠幅4.0/4.8 m，树龄139年，花期2-4月，长势良好。管护单位重庆市南山植物园管理处。

Located at the Guchayiyun, Chongqing Nanshan Botanical Garden, with latitude 29°33'25" N, longitude 106°37'19" E, elevation 535.7 m, tree H 4.8 m, two branches (43.0 cm/27.3 cm), CD 4.0 m/4.8 m, 139 y, Fl. Feb. to Apr. Grows well. Managed and maintained by Chongqing Nanshan Botanical Garden Administrative Office.

古38：白洋片
C. japonica 'Baiyangpian'

位于重庆市南山植物园古茶遗韵处，N29°33′25″，E106°37′19″，海拔535.5 m。株高4.0 m，基围66.6 cm，冠幅5.1/4.6 m，树龄112年，花期2–4月，长势良好。管护单位重庆市南山植物园管理处。

Located at the Guchayiyun, Chongqing Nanshan Botanical Garden, with latitude 29°33′25″ N, longitude 106°37′19″ E, elevation 535.5 m, tree H 4.0 m, CG 66.6 cm, CD 5.1 m/4.6 m, 112 y, Fl. Feb. to Apr. Grows well. Managed and maintained by Chongqing Nanshan Botanical Garden Administrative Office.

古39：紫金冠
C. japonica 'Zijinguan'

位于重庆市南山植物园花中魁首处，N29°33′25″，E106°37′19″，海拔535.4 m。株高5.0 m，基围90.4 cm，冠幅5.4/4.9 m，树龄216年，花期1–4月，长势良好。管护单位重庆市南山植物园管理处。

Located at the Huazhongkuishou, Chongqing Nanshan Botanical Garden, with latitude 29°33′25″ N, longitude 106°37′19″ E, elevation 535.4 m, tree H 5.0 m, CG 90.4 cm, CD 5.4 m/4.9 m, 216 y, Fl. Jan. to Apr. Grows well. Managed and maintained by Chongqing Nanshan Botanical Garden Administrative Office.

古 40：金顶大红
C. japonica 'Jinding Dahong'

位于重庆市南山植物园花中魁首处，N29°33′25″，E106°37′19″，海拔 535.5 m。株高 3.2 m，基围 85.1 cm，冠幅 5.5/7.0 m，树龄 203 年，花期 1–4 月，长势良好。管护单位重庆市南山植物园管理处。

Located at the Huazhongkuishou, Chongqing Nanshan Botanical Garden, with latitude 29°33′25″ N, longitude 106°37′19″ E, elevation 535.5 m, tree H 3.2 m, CG 85.1 cm, CD 5.5 m/7.0 m, 203 y, Fl. Jan. to Apr. Grows well. Managed and maintained by Chongqing Nanshan Botanical Garden Administrative Office.

古 41：紫金冠
C. japonica 'Zijinguan'

位于重庆市南山植物园花中魁首处，N29°33′24″，E106°37′19″，海拔 536.3 m。株高 6.8 m，基围 100.5 cm，冠幅 8.3/7.3 m，树龄 258 年，花期 1-4 月，长势良好。管护单位重庆市南山植物园管理处。

Located at the Huazhongkuishou, Chongqing Nanshan Botanical Garden, with latitude 29°33′24″ N, longitude 106°37′19″ E, elevation 536.3 m, tree H 6.8 m, CG 100.5 cm, CD 8.3 m/7.3 m, 258 y, Fl. Jan. to Apr. Grows well. Managed and maintained by Chongqing Nanshan Botanical Garden Administrative Office.

古42：白洋片
C. japonica 'Baiyangpian'

位于重庆市南山植物园花中魁首处，N29°33′24″，E106°37′18″，海拔535.6 m。株高4.0 m，基围65.3 cm，冠幅5.4/5.0 m，树龄112年，花期2–4月，长势良好。管护单位重庆市南山植物园管理处。

Located at the Huazhongkuishou, Chongqing Nanshan Botanical Garden, with latitude 29°33′24″ N, longitude 106°37′18″ E, elevation 535.6 m, tree H 4.0 m, CG 65.3 cm, CD 5.4 m/5.0 m, 112 y, Fl. Feb. to Apr. Grows well. Managed and maintained by Chongqing Nanshan Botanical Garden Administrative Office.

古 43: 红佛鼎
C. japonica 'Hongfoding'

位于重庆市南山植物园宏建山庄大门外山边堡坎上，N29°33′25″，E106°37′17″，海拔 540.0 m。株高 5.3 m，基围 57.8 cm，冠幅 4.7/5.3 m，树龄 152 年，花期 2-4 月，长势良好。管护单位重庆市南山植物园管理处。

Located on the hill outside the door of the Hongjian Villa, Chongqing Nanshan Botanical Garden, with latitude 29°33′25″ N, longitude 106°37′17″ E, elevation 540.0 m, tree H 5.3 m, CG 57.8 cm, CD 4.7 m/5.3 m, 152 y, Fl. Feb. to Apr. Grows well. Managed and maintained by Chongqing Nanshan Botanical Garden Administrative Office.

古44：紫金冠
C. japonica 'Zijinguan'

位于重庆市南山植物园宏建山庄大门外，N29°33′25″，E106°37′17″，海拔540.5 m。株高4.0 m，基围58.1 cm，冠幅3.8/4.6 m，树龄152年，花期1—4月。2016年雪灾造成树冠主枝受损，长势良好。管护单位重庆市南山植物园管理处。

Located outside the door of the Hongjian Villa, Chongqing Nanshan Botanical Garden, with latitude 29°33′25″ N, longitude 106°37′17″ E, elevation 540.5 m, tree H 4.0 m, CG 58.1 cm, CD 3.8 m/4.6 m, 152 y, Fl. Jan. to Apr. The branch was damaged by snow in 2016. Grows well. Managed and maintained by Chongqing Nanshan Botanical Garden Administrative Office.

古 45：七心白
C. japonica 'Qixinbai'

位于南山植物园丹霞亭前草坪上，N29°33′24″，E106°37′21″，海拔 523.3 m。株高 3.5 m，基围 61.9 cm，冠幅 4.0/4.3 m，树龄 148 年，花期 2–4 月，长势良好。管护单位重庆市南山植物园管理处。

Located at the lawn in front of the Danxia Pavilion, Chongqing Nanshan Botanical Garden, with latitude 29°33′24″ N, longitude 106°37′21″ E, elevation 523.3 m, tree H 3.5 m, CG 61.9 cm, CD 4.0 m/4.3 m, 148 y, Fl. Feb. to Apr. Grows well. Managed and maintained by Chongqing Nanshan Botanical Garden Administrative Office.

古46：金顶大红
C. japonica 'Jinding Dahong'

位于重庆市南山植物园川茶区丹霞亭处，N29°33′24″，E106°37′21″，海拔521.5 m。株高2.8 m，基围53.1 cm，冠幅4.2 m，树龄113年，花期1–4月，长势良好。管护单位重庆市南山植物园管理处。

Located at the Danxia Pavilion of the Sichuan Camellia Area, Chongqing Nanshan Botanical Garden, with latitude 29°33′24″ N, longitude 106°37′21″ E, elevation 521.5 m, tree H 2.8 m, CG 53.1 cm, CD 4.2 m, 113 y, Fl. Jan. to Apr. Grows well. Managed and maintained by Chongqing Nanshan Botanical Garden Administrative Office.

古 47：绒团茶
C. japonica 'Rongtuancha'

位于重庆市南山植物园川茶区草坪，N29°33′25″，E106°37′21″，海拔 525.3 m。株高 5.0 m，基围 57.5 cm，冠幅 6.0/5.0 m，树龄 107 年，花期 1–4 月，长势良好。管护单位重庆市南山植物园管理处。

Located at the lawn of the Sichuan Camellia Area, Chongqing Nanshan Botanical Garden, with latitude 29°33′25″ N, longitude 106°37′21″ E, elevation 525.3 m, tree H 5.0 m, CG 57.5 cm, CD 6.0 m/5.0 m, 107 y, Fl. Jan. to Apr. Grows well. Managed and maintained by Chongqing Nanshan Botanical Garden Administrative Office.

古 48：石榴茶
C. japonica 'Shiliu Cha'

位于重庆市南山植物园市花园石刻处，N29°33′23″，E106°37′22″，海拔 516.2 m。株高 6.5 m，两分枝 41.4/30.8 cm，冠幅 6.6/6.0 m，树龄 139 年，花期 1–4 月，长势良好。管护单位重庆市南山植物园管理处。

Located at the stone carving "Shihuayuan", Chongqing Nanshan Botanical Garden, with latitude 29°33′23″ N, longitude 106°37′22″ E, elevation 516.2 m, tree H 6.5 m, two branches (41.4 cm/30.8 cm), CD 6.6 m/6.0 m, 139 y, Fl. Jan. to Apr. Grows well. Managed and maintained by Chongqing Nanshan Botanical Garden Administrative Office.

古49：石榴茶
C. japonica 'Shiliu Cha'

位于重庆市南山植物园市花园石刻处，N29°33′23″，E106°37′22″，海拔516.4 m。株高4.5 m，基围55.9 cm，冠幅6.7 m，树龄193年，花期1–4月，长势良好。管护单位重庆市南山植物园管理处。

Located at the stone carving "Shihuayuan", Chongqing Nanshan Botanical Garden, with latitude 29°33′23″ N, longitude 106°37′22″ E, elevation 516.4 m, tree H 4.5 m, CG 55.9 cm, CD 6.7 m, 193 y, Fl. Jan. to Apr. Grows well. Managed and maintained by Chongqing Nanshan Botanical Garden Administrative Office.

古50：金顶大红
C. japonica 'Jinding Dahong'

位于重庆市南山植物园兰园冰井处，N29°33′26″，E106°37′23″，海拔518.3 m。株高3.5 m，两分枝46.8/27.9 cm，冠幅4.6/5.4 m，树龄139年，花期2-4月，长势良好。管护单位重庆市南山植物园管理处。

Located at the Ice Well of the Orchid Garden, Chongqing Nanshan Botanical Garden, with latitude 29°33′26″ N, longitude 106°37′23″ E, elevation 518.3 m, tree H 3.5m, two branches (46.8 cm/27.9 cm), CD 4.6 m/5.4 m, 139 y, Fl. Feb. to Apr. Grows well. Managed and maintained by Chongqing Nanshan Botanical Garden Administrative Office.

古51：重庆红
C. japonica 'Chongqinghong'

位于重庆市南山植物园兰园冰井处，N29°33′25″，E106°37′22″，海拔519.6 m。株高3.5 m，基围70.3 cm，冠幅4.3 m，树龄149年。由于之前所处环境植被太茂密，影响其生长，已移走周边植物，做了复壮措施。花期3–5月，长势一般。管护单位重庆市南山植物园管理处。

Located at the Ice Well of the Orchid Garden, Chongqing Nanshan Botanical Garden, with latitude 29°33′25″ N, longitude 106°37′22″ E, elevation 519.6 m, tree H 3.5 m, CG 70.3 cm, CD 4.3 m, 149 y. Growth was limited by previously surrounding environment. Surrounding plants were removed and rejuvenation measures were done. Fl. Mar. to May. Managed and maintained by Chongqing Nanshan Botanical Garden Administrative Office.

古52：重庆红
C. japonica 'Chongqinghong'

位于重庆市南山植物园兰溪水池边，N29°33′25″，E106°37′21″，海拔527.6 m。株高2.5 m，基围55.9 cm，冠幅4.8 m，树龄149年，花期3–5月，长势良好。管护单位重庆市南山植物园管理处。

Located beside the Lanxi Pool, Chongqing Nanshan Botanical Garden, with latitude 29°33′25″ N, longitude 106°37′21″ E, elevation 527.6 m, tree H 2.5 m, CG 55.9 cm, CD 4.8 m, 149 y. Fl. Mar. to May. Grows well. Managed and maintained by Chongqing Nanshan Botanical Garden Administrative Office.

古53: 白洋片
C. japonica 'Baiyangpian'

位于重庆市南山植物园西班牙公使馆前，N29°33'27″, E106°37'21″, 海拔535.2 m。株高5.0 m, 基围59.3 cm, 冠幅4.6 m, 树龄112年, 花期2–4月, 长势良好。管护单位重庆市南山植物园管理处。

Located in front of the Spanish mansion, Chongqing Nanshan Botanical Garden, with latitude 29°33'27″ N, longitude 106°37'21″ E, elevation 535.2 m, tree H 5.0 m, CG 59.3 cm, CD 4.6 m, 112 y, Fl. Feb. to Apr. Grows well. Managed and maintained by Chongqing Nanshan Botanical Garden Administrative Office.

古 54：醉杨妃
C. japonica 'Zuiyangfei'

位于重庆市南山植物园西班牙公使馆前，N29°33′27″，E106°37′20″，海拔 535.4 m。株高 4.0 m，基围 82.9 cm，冠幅 6.0 m，树龄 212 年，花期 2–4 月，是目前已知该品种中树龄最长的一棵古树，长势一般，已列入重点管护。管护单位重庆市南山植物园管理处。

Located in front of the Spanish mansion, Chongqing Nanshan Botanical Garden, with latitude 29°33′27″ N, longitude 106°37′20″ E, elevation 535.4 m, tree H 4.0 m, CG 82.9 cm, CD 6.0 m, 212 y, Fl. Feb. to Apr. It is the oldest known tree of this cultivar and is under focused management. Managed and maintained by Chongqing Nanshan Botanical Garden Administrative Office.

古55：紫金冠
C. japonica 'Zijinguan'

位于重庆市南山植物园图书馆往八德亭的路边上，N29°33′28″，E106°37′22″，海拔536.9 m。株高7.5 m，基围61.5 cm，冠幅4.9 m，树龄152年，花期1–4月，长势良好。管护单位重庆市南山植物园管理处。

Located at the roadside from library to Bade Pavilion, Chongqing Nanshan Botanical Garden, with latitude 29°33′28″ N, longitude 106°37′22″ E, elevation 536.9 m, tree H 7.5 m, CG 61.5 cm, CD 4.9 m, 152 y, Fl. Jan. to Apr. Grows well. Managed and maintained by Chongqing Nanshan Botanical Garden Administrative Office.

古56：紫金冠
C. japonica 'Zijinguan'

位于重庆市南山植物园图书馆往八德亭的路边上，N29°33′28″，E106°37′22″，海拔536.7 m。株高7.5 m，基围72.2 cm，冠幅4.6 m，树龄152年，花期1–4月，长势良好。管护单位重庆市南山植物园管理处。

Located at the roadside from library to Bade Pavilion, Chongqing Nanshan Botanical Garden, with latitude 29°33′28″ N, longitude 106°37′22″ E, elevation 536.7 m, tree H 7.5 m, CG 72.2 cm, CD 4.6 m, 152 y, Fl. Jan. to Apr. Grows well. Managed and maintained by Chongqing Nanshan Botanical Garden Administrative Office.

古 57：紫金冠
C. japonica 'Zijinguan'

位于重庆市南山植物园图书馆往八德亭的路边上，N29°33′28″，E106°37′22″，海拔536.5 m。株高7.0 m，基围53.4 cm，冠幅3.5 m，树龄152年，花期1-4月，长势良好。管护单位重庆市南山植物园管理处。

Located at the roadside from library to Bade Pavilion, Chongqing Nanshan Botanical Garden, with latitude 29°33′28″ N, longitude 106°37′22″ E, elevation 536.5 m, tree H 7.0 m, CG 53.4 cm, CD 3.5m, 152 y, Fl. Jan. to Apr. Grows well. Managed and maintained by Chongqing Nanshan Botanical Garden Administrative Office.

古58：紫金冠
C. japonica 'Zijinguan'

位于重庆市南山植物园图书馆往八德亭的路边上，N29°33′28″，E106°37′22″，海拔536.3 m。株高6.0 m，基围61.2 cm，冠幅3.3 m，树龄152年，花期1–4月，长势良好。管护单位重庆市南山植物园管理处。

Located at the roadside from library to Bade Pavilion, Chongqing Nanshan Botanical Garden, with latitude 29°33′28″ N, longitude 106°37′22″ E, elevation 536.3 m, tree H 6.0 m, CG 61.2 cm, CD 3.3 m, 152 y, Fl. Jan. to Apr. Grows well. Managed and maintained by Chongqing Nanshan Botanical Garden Administrative Office.

古59：紫金冠
C. japonica 'Zijinguan'

位于重庆市南山植物园图书馆往八德亭的路边上，N29°33′28″，E106°37′22″，海拔536.1 m。株高5.0 m，基围58.1 cm，冠幅3.3 m，树龄152年，花期1–4月，长势良好。管护单位重庆市南山植物园管理处。

Located at the roadside from library to Bade Pavilion, Chongqing Nanshan Botanical Garden, with latitude 29°33′28″ N, longitude 106°37′22″ E, elevation 536.1 m, tree H 5.0 m, CG 58.1 cm, CD 3.3 m, 152 y, Fl. Jan. to Apr. Grows well. Managed and maintained by Chongqing Nanshan Botanical Garden Administrative Office.

古60：紫金冠
C. japonica 'Zijinguan'

位于重庆市南山植物园图书馆往八德亭的路边上，N29°33′28″，E106°37′23″，海拔535.3 m。株高6.0 m，基围59.7 cm，冠幅4.4 m，树龄152年，花期1–4月，长势良好。管护单位重庆市南山植物园管理处。

Located at the roadside from library to Bade Pavilion, Chongqing Nanshan Botanical Garden, with latitude 29°33′28″ N, longitude 106°37′23″ E, elevation 535.3 m, tree H 6.0 m, CG 59.7 cm, CD 4.4 m, 152 y, Fl. Jan. to Apr. Grows well. Managed and maintained by Chongqing Nanshan Botanical Garden Administrative Office.

古61：紫金冠
C. japonica 'Zijinguan'

位于重庆市南山植物园图书馆往八德亭的路边上，N29°33′29″，E106°37′23″，海拔534.9 m。株高6.5 m，三分枝33.0/41.8/50.2 cm，冠幅5.2 m，树龄152年，花期1-4月，长势良好。管护单位重庆市南山植物园管理处。

Located at the roadside from library to Bade Pavilion, Chongqing Nanshan Botanical Garden, with latitude 29°33′29″ N, longitude 106°37′23″ E, elevation 534.9 m, tree H 6.5 m, three branches (33.0 cm/41.8 cm/50.2 cm), CD 5.2 m, 152 y, Fl. Jan. to Apr. Grows well. Managed and maintained by Chongqing Nanshan Botanical Garden Administrative Office.

古 62: 白洋片
C. japonica 'Baiyangpian'

位于重庆市南山植物园图书馆往八德亭的路边上，N29°33′29″，E106°37′23″，海拔535.2 m。株高4.7 m，基围57.5 cm，冠幅3.8 m，树龄152年，花期1–4月，长势良好。管护单位重庆市南山植物园管理处。

Located at the roadside from library to Bade Pavilion, Chongqing Nanshan Botanical Garden, with latitude 29°33′29″ N, longitude 106°37′23″ E, elevation 535.2 m, tree H 4.7 m, CG 57.5 cm, CD 3.8 m, 152 y, Fl. Jan. to Apr. Grows well. Managed and maintained by Chongqing Nanshan Botanical Garden Administrative Office.

古63：紫金冠
C. japonica 'Zijinguan'

位于重庆市南山植物园图书馆往八德亭的路边上，N29°33′29″，E106°37′23″，海拔535.7 m。株高5.0 m，基围46.2 cm，冠幅4.3/5.0 m，树龄152年，花期1-4月，长势良好。管护单位重庆市南山植物园管理处。

Located at the roadside from library to Bade Pavilion, Chongqing Nanshan Botanical Garden, with latitude 29°33′29″ N, longitude 106°37′23″ E, elevation 535.7 m, tree H 5.0 m, CG 46.2 cm, CD 4.3 m/5.0 m, 152 y, Fl. Jan. to Apr. Grows well. Managed and maintained by Chongqing Nanshan Botanical Garden Administrative Office.

古64: 紫金冠
C. japonica 'Zijinguan'

位于重庆市南山植物园八德亭下路边,N29°33′29″,E106°37′24″,海拔531.3 m。株高5.0 m,基围65.9 cm,冠幅3.4/3.1 m,树龄152年,花期1–4月,长势良好。管护单位重庆市南山植物园管理处。

Located at the roadside of the Bade Pavilion, Chongqing Nanshan Botanical Garden, with latitude 29°33′29″ N, longitude 106°37′24″ E, elevation 531.3 m, tree H 5.0 m, CG 65.9 cm, CD 3.4 m/3.1 m, 152 y, Fl. Jan. to Apr. Grows well. Managed and maintained by Chongqing Nanshan Botanical Garden Administrative Office.

古65：紫金冠
C. japonica 'Zijinguan'

位于重庆市南山植物园八德亭处，N29°33′28″，E106°37′24″，海拔531.0 m。株高4.7 m，基围54.0 cm，冠幅3.7/3.2 m，树龄152年，花期1–4月，长势良好。管护单位重庆市南山植物园管理处。

Located at the Bade Pavilion of Chongqing Nanshan Botanical Garden, with latitude 29°33′28″ N, longitude 106°37′24″ E, elevation 531.0 m, tree H 4.7 m, CG 54.0 cm, CD 3.7 m/3.2 m, 152 y, Fl. Jan. to Apr. Grows well. Managed and maintained by Chongqing Nanshan Botanical Garden Administrative Office.

古66：紫金冠
C. japonica 'Zijinguan'

位于重庆市南山植物园八德亭处，N29°33′28″，E106°37′24″，海拔530.4 m。株高6.0 m，基围59.0 cm，冠幅5.0/4.0 m，树龄152年，花期1–4月，长势良好。管护单位重庆市南山植物园管理处。

Located at the Bade Pavilion of Chongqing Nanshan Botanical Garden, with latitude 29°33′28″ N, longitude 106°37′24″ E, elevation 530.4 m, tree H 6.0 m, CG 59.0 cm, CD 5.0 m/4.0 m, 152 y, Fl. Jan. to Apr. Grows well. Managed and maintained by Chongqing Nanshan Botanical Garden Administrative Office.

古 67：紫金冠
C. japonica 'Zijinguan'

位于重庆市南山植物园八德亭处，N29°33′28″，E106°37′24″，海拔 531.1 m。株高 6.5 m，基围 69.1 cm，冠幅 5.5/4.9 m，树龄 152 年，花期 1–4 月，长势良好。管护单位重庆市南山植物园管理处。

Located at the Bade Pavilion of Chongqing Nanshan Botanical Garden, with latitude 29°33′28″ N, longitude 106°37′24″ E, elevation 531.1 m, tree H 6.5 m, CG 69.1 cm, CD 5.5 m/4.9 m, 152 y, Fl. Jan. to Apr. Grows well. Managed and maintained by Chongqing Nanshan Botanical Garden Administrative Office.

古68：紫金冠
C. japonica 'Zijinguan'

位于重庆市南山植物园八德亭处，N29°33′28″，E106°37′24″，海拔 527.3 m。株高 6.5 m，基围 59.7 cm，冠幅 5.0/3.7 m，树龄 152 年，花期 1–4 月，长势良好。管护单位重庆市南山植物园管理处。

Located at the Bade Pavilion of Chongqing Nanshan Botanical Garden, with latitude 29°33′28″ N, longitude 106°37′24″ E, elevation 527.3 m, tree H 6.5 m, CG 59.7 cm, CD 5.0 m/3.7 m, 152 y, Fl. Jan. to Apr. Grows well. Managed and maintained by Chongqing Nanshan Botanical Garden Administrative Office.

古 69：紫金冠
C. japonica 'Zijinguan'

位于重庆市南山植物园八德亭处，N29°33′28″，E106°37′24″，海拔 526.7 m。株高 6.0 m，基围 60.9 cm，冠幅 5.8/4.7 m，树龄 152 年，花期 1–4 月，长势良好。管护单位重庆市南山植物园管理处。

Located at the Bade Pavilion of Chongqing Nanshan Botanical Garden, with latitude 29°33′28″ N, longitude 106°37′24″ E, elevation 526.7 m, tree H 6.0 m, CG 60.9 cm, CD 5.8 m/4.7 m, 152 y, Fl. Jan. to Apr. Grows well. Managed and maintained by Chongqing Nanshan Botanical Garden Administrative Office.

古70：紫金冠
C. japonica 'Zijinguan'

位于重庆市南山植物园八德亭处，N29°33′28″，E106°37′24″，海拔528.7 m。株高6.7 m，基围76.6 cm，冠幅6.0/6.3 m，树龄152年，花期1–4月，长势良好。管护单位重庆市南山植物园管理处。

Located at the Bade Pavilion of Chongqing Nanshan Botanical Garden, with latitude 29°33′28″ N, longitude 106°37′24″ E, elevation 528.7 m, tree H 6.7 m, CG 76.6 cm, CD 6.0 m/6.3 m, 152 y, Fl. Jan. to Apr. Grows well. Managed and maintained by Chongqing Nanshan Botanical Garden Administrative Office.

古71：紫金冠
C. japonica 'Zijinguan'

位于重庆市南山植物园八德亭处，N29°33′28″，E106°37′24″，海拔 526.2 m。株高 6.3 m，三分枝 26.1/33.9/54.3 cm，冠幅 6.0 m，树龄 152 年，花期 1–4 月，长势良好。管护单位重庆市南山植物园管理处。

Located at the Bade Pavilion of Chongqing Nanshan Botanical Garden, with latitude 29°33′28″ N, longitude 106°37′24″ E, elevation 526.2 m, tree H 6.3 m, three branches (26.1 cm/33.9 cm/54.3 cm), CD 6.0 m, 152 y, Fl. Jan. to Apr. Grows well. Managed and maintained by Chongqing Nanshan Botanical Garden Administrative Office.

古72：紫金冠
C. japonica 'Zijinguan'

位于重庆市南山植物园八德亭处，N29°33′28″，E106°37′24″，海拔525.4 m，株高5.8 m，基围58.4 cm，冠幅5.6/4.8 m，树龄152年，花期1–4月，长势良好。管护单位重庆市南山植物园管理处。

Located at the Bade Pavilion of Chongqing Nanshan Botanical Garden, with latitude 29°33′28″ N, longitude 106°37′24″ E, elevation 525.4 m, tree H 5.8 m, CG 58.4 cm, CD 5.6 m/4.8 m, 152 y. Fl. Jan. to Apr. Grows well. Managed and maintained by Chongqing Nanshan Botanical Garden Administrative Office.

古 73：紫金冠
C. japonica 'Zijinguan'

位于重庆市南山植物园八德亭处，N29°33′28″，E106°37′25″，海拔 524.3 m。株高 5.8 m，四分枝 27.3/41.4/17.6/23.6 cm，冠幅 6.0/4.5 m，树龄 152 年，花期 1–4 月，长势良好。管护单位重庆市南山植物园管理处。

Located at the Bade Pavilion of Chongqing Nanshan Botanical Garden, with latitude 29°33′28″ N, longitude 106°37′25″ E, elevation 524.3 m, tree H 5.8 m, four branches (27.3 cm/41.4 cm/17.6 cm/23.6 cm), CD 6.0 m/4.5 m, 152 y, Fl. Jan. to Apr. Grows well. Managed and maintained by Chongqing Nanshan Botanical Garden Administrative Office.

古 74：紫金冠
C. japonica 'Zijinguan'

位于重庆市南山植物园八德亭处，N29°33′28″，E106°37′25″，海拔 524.6 m。株高 5.3 m，基围 66.6 cm，冠幅 4.3 m，树龄 152 年，花期 1–4 月，长势良好。管护单位重庆市南山植物园管理处。

Located at the Bade Pavilion of Chongqing Nanshan Botanical Garden, with latitude 29°33′28″ N, longitude 106°37′25″ E, elevation 524.6 m, tree H 5.3 m, CG 66.6 cm, CD 4.3 m, 152 y, Fl. Jan. to Apr. Grows well. Managed and maintained by Chongqing Nanshan Botanical Garden Administrative Office.

古75：紫金冠
C. japonica 'Zijinguan'

位于重庆市南山植物园八德亭处，N29°33′28″，E106°37′25″，海拔526.0 m。株高5.0 m，三分枝38.6/20.4/51.8 cm，冠幅4.8 m，树龄152年，花期1–4月，长势良好。管护单位重庆市南山植物园管理处。

Located at the Bade Pavilion of Chongqing Nanshan Botanical Garden, with latitude 29°33′28″ N, longitude 106°37′25″ E, elevation 526.0 m, tree H 5.0 m, three branches (38.6 cm/20.4 cm/51.8 cm), CD 4.8 m, 152 y, Fl. Jan. to Apr. Grows well. Managed and maintained by Chongqing Nanshan Botanical Garden Administrative Office.

古76：白洋片
C. japonica 'Baiyangpian'

位于重庆市南山植物园八德亭往山茶多样性展示区路边，N29°33′29″，E106°37′24″，海拔526.2 m。株高3.3 m，基围44.0 cm，冠幅2.8 m，树龄106年，花期2–4月，长势良好。管护单位重庆市南山植物园管理处。

Located at the road side from the Bade Pavilion to the Camellia Diversity Area, Chongqing Nanshan Botanical Garden, with latitude 29°33′29″ N, longitude 106°37′24″ E, elevation 526.2 m, tree H 3.3 m, CG 44.0 cm, CD 2.8 m, 106 y, Fl. Feb. to Apr. Grows well. Managed and maintained by Chongqing Nanshan Botanical Garden Administrative Office.

古 77: 三学士
C. japonica 'Sanxueshi'

位于重庆市南山植物园山茶多样性展示区路口处，N29°33′29″，E106°37′24″，海拔525.4 m。株高3.8 m，基围44.9 cm，冠幅4.2/3.3 m，树龄104年，花期3–5月，长势良好。管护单位重庆市南山植物园管理处。

Located at the crossroads of the Camellia Diversity Area, Chongqing Nanshan Botanical Garden, with latitude 29°33′29″ N, longitude 106°37′24″ E, elevation 525.4 m, tree H 3.8 m, CG 44.9 cm, CD 4.2 m/3.3 m, 104 y, Fl. Mar. to May. Grows well. Managed and maintained by Chongqing Nanshan Botanical Garden Administrative Office.

古 78：醉杨妃
C. japonica 'Zuiyangfei'

位于重庆市南山植物园山茶园多样性展示区路边，N29°33′30″，E106°37′25″，海拔 522.0 m。株高 4.3 m，基围 44.0 cm，树干两分枝，其中一枝空腐，只剩下树皮。冠幅 3.4/2.7 m，树龄 112 年，花期 2–4 月，长势良好。管护单位重庆市南山植物园管理处。

Located at the roadside of the Camellia Diversity Area, Chongqing Nanshan Botanical Garden, with latitude 29°33′30″ N, longitude 106°37′25″ E, elevation 522.0 m, tree H 4.3 m, CG 44.0 cm, two branches but one is rotted. CD 3.4 m/2.7 m, 112 y, Fl. Feb. to Apr. Grows well. Managed and maintained by Chongqing Nanshan Botanical Garden Administrative Office.

古79：七心白
C. japonica 'Qixinbai'

位于重庆市南山植物园山茶园多样性展示区路边，N29°33'30″，E106°37'25″，海拔520.9 m。株高2.8 m，基围49.6 cm，冠幅2.7/3.0 m，树龄148年，花期2–4月，长势良好。管护单位重庆市南山植物园管理处。

Located at the roadside of the Camellia Diversity Area, Chongqing Nanshan Botanical Garden, with latitude 29°33'30″ N, longitude 106°37'25″ E, elevation 520.9 m, tree H 2.8 m, CG 49.6 cm, CD 2.7 m/3.0 m, 148 y, Fl. Feb. to Apr. Grows well. Managed and maintained by Chongqing Nanshan Botanical Garden Administrative Office.

古80：紫金冠
C. japonica 'Zijinguan'

位于重庆市南山植物园紫金冠大道，N29°33′28″，E106°37′25″，海拔521.1 m。株高7.0 m，基围50.2 cm，冠幅5.0 m，树龄112年，花期1–4月，长势良好。营护单位重庆市南山植物园管理处。

Located at the side of the Zijinguan Road, Chongqing Nanshan Botanical Garden, with latitude 29°33′28″ N, longitude 106°37′25″ E, elevation 521.1 m, tree H 7.0 m, CG 50.2 cm, CD 5.0 m, 112 y, Fl. Jan. to Apr. Grows well. Managed and maintained by Chongqing Nanshan Botanical Garden Administrative Office.

古 81: 紫金冠
C. japonica 'Zijinguan'

位于重庆市南山植物园蔷薇园白兰花广场，N29°33′27″，E106°37′42″，海拔 458.9 m。株高 3.9 m，基围 57.5 cm，冠幅 5.0/5.4 m，树龄 152 年，花期 1–4 月，长势良好。据史料记载，南山植物园蔷薇园内多数茶花古树是原民国实业家范崇实私家花园保留下来的。管护单位重庆市南山植物园管理处。

Located in the White Orchid Plaza of the Rose Garden, Chongqing Nanshan Botanical Garden, with latitude 29°33′27″ N, longitude 106°37′42″ E, elevation 458.9 m, tree H 3.9 m, CG 57.5 cm, CD 5.0 m/5.4 m, 152 y, Fl. Jan. to Apr. Grows well. According to historical records, majority of ancient camellia trees in the Rose Garden were preserved from the private garden of Fan Chongshi, an industrialist from the Republic of China. Managed and maintained by Chongqing Nanshan Botanical Garden Administrative Office.

古 82：七心红
C. japonica 'Qixinhong'

位于重庆市南山植物园蔷薇园白兰花广场，N29°33'27"，E106°37'43"，海拔 459.0 m。株高 3.5 m，基围 82.3 cm，冠幅 4.6/6.0 m，树龄 172 年，花期 1–4 月，长势良好。管护单位重庆市南山植物园管理处。

Located in the White Orchid Plaza of the Rose Garden, Chongqing Nanshan Botanical Garden, with latitude 29°33'27" N, longitude 106°37'43" E, elevation 459.0 m, tree H 3.5 m, CG 82.3 cm, CD 4.6 m/6.0 m, 172 y, Fl. Jan. to Apr. Grows well. Managed and maintained by Chongqing Nanshan Botanical Garden Administrative Office.

古 83: 白洋片
C. japonica 'Baiyangpian'

位于重庆市南山植物园蔷薇园白兰花广场，N29°33′27″，E106°37′43″，海拔 453.4 m。株高 3.6 m，基围 76.9 cm，冠幅 5.0 m，树龄 149 年，花期 1–4 月，长势良好。管护单位重庆市南山植物园管理处。

Located in the White Orchid Plaza of the Rose Garden, Chongqing Nanshan Botanical Garden, with latitude 29°33′27″ N, longitude 106°37′43″ E, elevation 453.4 m, tree H 3.6 m, CG 76.9 cm, CD 5.0 m, 149 y, Fl. Jan. to Apr. Grows well. Managed and maintained by Chongqing Nanshan Botanical Garden Administrative Office.

古84：氅盔
C. japonica 'Changkui'

位于重庆市南山植物园蔷薇园白兰花广场，N29°33′27″，E106°37′43″，海拔458.8 m。株高4.5 m，两分枝67.5/60.0 cm，冠幅6.5/7.5 m，树龄188年，花期1—4月，长势良好。管护单位重庆市南山植物园管理处。

Located in the White Orchid Plaza of the Rose Garden, Chongqing Nanshan Botanical Garden, with latitude 29°33′27″ N, longitude 106°37′43″ E, elevation 458.8 m, tree H 4.5 m, two branches (67.5 cm/60.0 cm), CD 6.5 m/7.5 m, 188 y, Fl. Jan. to Apr. Grows well. Managed and maintained by Chongqing Nanshan Botanical Garden Administrative Office.

古85：七心红
C. japonica 'Qixinhong'

位于重庆市南山植物园蔷薇园白兰花广场，N29°33′27″，E106°37′43″，海拔459.0 m。株高2.8 m，基围44.9 cm，冠幅6.0/4.5 m，树龄113年，花期1–4月。因树冠过大，枝叶茂密，为防止折断，已进行支撑加固处理，长势良好。管护单位重庆市南山植物园管理处。

Located in the White Orchid Plaza of the Rose Garden, Chongqing Nanshan Botanical Garden, with latitude 29°33′27″ N, longitude 106°37′43″ E, elevation 459.0 m, tree H 2.8 m, CG 44.9 cm , CD 6.0 m/4.5 m, 113 y, Fl. Jan. to Apr. To prevent undesired break due to giant canopy and dense branches, supporting and reinforcing measures have been done. Grows well. Managed and maintained by Chongqing Nanshan Botanical Garden Administrative Office.

古 86：白洋片
C. japonica 'Baiyangpian'

位于重庆市南山植物园蔷薇园花神雕塑处，N29°33′28″，E106°37′43″，海拔 460.9 m。株高 3.5 m，基围 49.9 cm，冠幅 4.1 m，树龄 106 年，花期 2–4 月，长势良好。管护单位重庆市南山植物园管理处。

Located at the sculpture "Huashen" of the Rose Garden, Chongqing Nanshan Botanical Garden, with latitude 29°33′28″ N, longitude 106°37′43″ E, elevation 460.9 m, tree H 3.5 m, CG 49.9 cm, CD 4.1 m, 106 y, Fl. Feb. to Apr. Grows well. Managed and maintained by Chongqing Nanshan Botanical Garden Administrative Office.

古87：七心红
C. japonica 'Qixinhong'

位于重庆市南山植物园蔷薇园花神雕塑处，N29°33′29″，E106°37′43″，海拔 460.8 m。株高 3.5 m，基围 59.7 cm，冠幅 5.8/4.5 m，树龄 113 年，花期 1–4 月，长势良好。管护单位重庆市南山植物园管理处。

Located at the sculpture "Huashen" of the Rose Garden, Chongqing Nanshan Botanical Garden, with latitude 29°33′29″ N, longitude 106°37′43″ E, elevation 460.8 m, tree H 3.5 m, CG 59.7 cm, CD 5.8 m/4.5 m, 113 y, Fl. Jan. to Apr. Grows well. Managed and maintained by Chongqing Nanshan Botanical Garden Administrative Office.

古 88：花洋红
C. japonica 'Hua Yanghong'

位于重庆市南山植物园蔷薇园花神雕塑处，N29°33′29″，E106°37′43″，海拔 461.3 m。株高 3.8 m，基围 55.6 cm，冠幅 3.0/2.6 m，树龄 107 年，花期 1–4 月，长势一般。管护单位重庆市南山植物园管理处。

Located at the sculpture "Huashen" of the Rose Garden, Chongqing Nanshan Botanical Garden, with latitude 29°33′29″ N, longitude 106°37′43″ E, elevation 461.3 m, tree H 3.8 m, CG 55.6 cm , CD 3.0 m/2.6 m, 107 y, Fl. Jan. to Apr. Managed and maintained by Chongqing Nanshan Botanical Garden Administrative Office.

古 89：抓破脸
C. japonica 'Zhuapolian'

位于重庆市南山植物园蔷薇园逍遥坡上，N29°33′28″, E 106°37′41″, 海拔 463.2 m。株高 3.5 m，基围 55.9 cm, 冠幅 4.4/5.9 m, 树龄 121 年，花期 2–4 月，长势良好。管护单位重庆市南山植物园管理处。

Located at the Xiaoyao hillside of the Rose Garden, Chongqing Nanshan Botanical Garden, with latitude 29°33′28″ N, longitude 106°37′41″ E, elevation 463.2 m, tree H 3.5 m, CG 55.9 cm, CD 4.4 m/5.9 m, 121 y, Fl. Feb. to Apr. Grows well. Managed and maintained by Chongqing Nanshan Botanical Garden Administrative Office.

古 90：石榴茶
C. japonica 'Shiliu Cha'

位于重庆市南山植物园逍遥坡上，N29°33′29″，E106°37′40″，海拔 470.6 m。株高 4.0 m，两分枝 38.6/36.7 cm，冠幅 7.0/6.0 m，树龄 139 年，花期 2–4 月，是范家花园保留下来的，长势良好。管护单位重庆市南山植物园管理处。

Located at the Xiaoyao hillside of the Rose Garden, Chongqing Nanshan Botanical Garden, with latitude 29°33′29″ N, longitude 106°37′40″ E, elevation 470.6 m, tree H 4.0 m, two branches (38.6 cm/36.7 cm), CD 7.0 m/6.0 m, 139 y, Fl. Feb. to Apr. Grows well. Preserved from the private garden of Fan Chongshi. Managed and maintained by Chongqing Nanshan Botanical Garden Administrative Office.

古91：紫金冠
C. japonica 'Zijinguan'

位于重庆市南山植物园蔷薇园逍遥坡上，N29°33′29″，E106°37′40″，海拔471.4 m。株高4.5 m，基围53.1 cm，冠幅4.6 m，树龄169年，花期1–4月，长势良好。管护单位重庆市南山植物园管理处。

Located at the Xiaoyao hillside of the Rose Garden, Chongqing Nanshan Botanical Garden, with latitude 29°33′29″ N, longitude 106°37′40″ E, elevation 471.4 m, tree H 4.5 m, CG 53.1 cm, CD 4.6 m, 169 y, Fl. Jan. to Apr. Grows well. Managed and maintained by Chongqing Nanshan Botanical Garden Administrative Office.

古 92：紫金冠
C. japonica 'Zijinguan'

位于重庆市南山植物园蔷薇园逍遥坡上，N29°33′30″，E106°37′40″，海拔 472.1 m。株高 4.9 m，两分枝 47.1/42.4 cm，冠幅 5.0/6.0 m，树龄 169 年，花期 1–4 月，长势良好。管护单位重庆市南山植物园管理处。

Located at the Xiaoyao hillside of the Rose Garden, Chongqing Nanshan Botanical Garden, with latitude 29°33′30″ N, longitude 106°37′40″ E, elevation 472.1 m, tree H 4.9 m, two branches (47.1 cm/42.4 cm), CD 5.0 m/6.0 m, 169 y, Fl. Jan. to Apr. Grows well. Managed and maintained by Chongqing Nanshan Botanical Garden Administrative Office.

古 93：紫金冠
C. japonica 'Zijinguan'

位于重庆市南山植物园蔷薇园逍遥坡上，N29°33′30″，E106°37′40″，海拔 470.0 m。株高 4.0 m，两分枝 42.1/36.4 cm，冠幅 4.1/4.7 m，树龄 169 年，花期 1–4 月，长势良好。管护单位重庆市南山植物园管理处。

Located at the Xiaoyao hillside of the Rose Garden, Chongqing Nanshan Botanical Garden, with latitude 29°33′30″ N, longitude 106°37′40″ E, elevation 470.0 m, tree H 4.0 m, two branches (42.1 cm/36.4 cm), CD 4.1 m/4.7 m, 169 y, Fl. Jan. to Apr. Grows well. Managed and maintained by Chongqing Nanshan Botanical Garden Administrative Office.

古 94：抓破脸
C. japonica 'Zhuapolian'

位于重庆市南山植物园蔷薇园海棠烟雨路下，N29°33′31″，E106°37′39″，海拔 470.4 m。株高 3.5 m，基围 58.1 cm，冠幅 4.2 m，树龄 121 年，花期 2–4 月，长势良好。管护单位重庆市南山植物园管理处。

Located under the Haitangyanyu Road, Chongqing Nanshan Botanical Garden, with latitude 29°33′31″ N, longitude 106°37′39″ E, elevation 470.4 m, tree H 3.5 m, CG 58.1 cm , CD 4.2 m, 121 y, Fl. Feb. to Apr. Grows well. Managed and maintained by Chongqing Nanshan Botanical Garden Administrative Office.

古95: 醉杨妃
C. japonica 'Zuiyangfei'

位于重庆市南山植物园山茶园贞寿亭处，N29°33′25″，E108°37′20″，海拔581.9 m。株高3.0 m，基围53.0 cm，三分枝，冠幅3.0 m，树龄112年，花期2-4月，长势良好。管护单位重庆市南山植物园管理处。

Located at the Zhenshou Pavilion of Chongqing Nanshan Botanical Garden, with latitude 29°33′25″ N, longitude 108°37′20″ E, elevation 581.9 m, tree H 3.0 m, CG 53.0 cm, three branches, CD 3.0 m, 112 y, Fl. Feb. to Apr. Grows well. Managed and maintained by Chongqing Nanshan Botanical Garden Administrative Office.

古 96：花洋红
C. japonica 'Hua Yanghong'

位于重庆市南山植物园山茶园贞寿亭车行道下方，N29°33′25″，E106°37′20″，海拔569.6 m。株高4.6 m，基围50.0 cm，冠幅4.0 m，两分枝，树龄107年，花期1~4月，长势良好。管护单位重庆市南山植物园管理处。

Located at the roadside closed to the Zhenshou Pavilion, Chongqing Nanshan Botanical Garden, with latitude 29°33′25″ N, longitude 106°37′20″ E, elevation 569.6 m, tree H 4.6 m, CG 50.0 cm, CD 4.0 m, two branches, 107 y, Fl. Jan. to Apr. Grows well. Managed and maintained by Chongqing Nanshan Botanical Garden Administrative Office.

古97：花洋红
C. japonica 'Hua Yanghong'

位于重庆市南山植物园兰园揽月波影前草坪上，N29°33′27″，E106°37′22″，海拔555.9 m。株高4.0 m，基围56.0 cm，冠幅4.2 m，两分枝，树龄123年，花期1–4月，长势良好。管护单位重庆市南山植物园管理处。

Located on the lawn in front of the Lanyueboying, Orchid Garden, Chongqing Nanshan Botanical Garden, with latitude 29°33′27″ N, longitude 106°37′22″ E, elevation 555.9 m, tree H 4.0 m, CG 56.0 cm, CD 4.2 m, two branches, 123 y, Fl. Jan. to Apr. Grows well. Managed and maintained by Chongqing Nanshan Botanical Garden Administrative Office.

古98：金顶大红
C. japonica 'Jinding Dahong'

位于重庆市南山植物园桂花山靠古茶苑路边上，N29°33′27″，E106°37′19″，海拔586.6 m。株高3.0 m，基围85.0 cm，冠幅4.0 m，两分枝，树龄113年，花期1-4月，长势良好。管护单位重庆市南山植物园管理处。

Located on the Osmanthus hill closed to the Gucha Yard, Chongqing Nanshan Botanical Garden, with latitude 29°33′27″ N, longitude 106°37′19″ E, elevation 586.6 m, tree H 3.0 m, CG 85.0 cm, CD 4.0 m, two branches, 113 y, Fl. Jan. to Apr. Grows well. Managed and maintained by Chongqing Nanshan Botanical Garden Administrative Office.

古99：金顶大红
C. japonica 'Jinding Dahong'

位于重庆市南山植物园桂花山上，N29°33′26″，E106°37′18″，海拔572.3 m。株高3.5 m，基围90.0 cm，三分枝，最大两分枝35.0/47.0 cm，冠幅5.0 m，树龄113年，花期1–4月，长势良好。管护单位重庆市南山植物园管理处。

Located on the Osmanthus hill, Chongqing Nanshan Botanical Garden, with latitude 29°33′26″ N, longitude 106°37′18″ E, elevation 572.3 m, tree H 3.5 m, CG 90.0 cm, three branches (the biggest two 35.0 cm/47.0 cm respectively), CD 5.0 m, 113 y, Fl. Jan. to Apr. Grows well. Managed and maintained by Chongqing Nanshan Botanical Garden Administrative Office.

古100：金顶大红
C. japonica 'Jinding Dahong'

位于重庆市南山植物园金鹰园入口处，N29°33′28″，E106°37′12″，海拔579.2 m。株高2.5 m，基围60.0 cm，两分枝35.0/33.0 cm，冠幅4.0 m，树龄139年，花期1–4月，树干多处空腐，长势良好。管护单位重庆市南山植物园管理处。

Located at the entrance of the Gold Eagle Garden, Chongqing Nanshan Botanical Garden, with latitude 29°33′28″ N, longitude 106°37′12″ E, elevation 579.2 m, tree H 2.5 m, CG 60.0 cm, two branches (35.0 cm/33.0 cm), CD 4.0 m, 139 y, Fl. Jan. to Apr. Grows well. The trunk is rotted at several places. Managed and maintained by Chongqing Nanshan Botanical Garden Administrative Office.

万州
西山公园

INTRODUCTION TO
WANZHOU XISHAN PARK

万州西山公园建于 1925 年，四川军阀杨森进驻万州，在原万州西山观的基础上改建为万州西山公园。万州西山公园是全国最早著名公园之一，也是全国著名古茶花园之一。园内林木苍翠，山茶栽于建园之初，经过近百年的精心培育，现有川山茶古树 110 株。2016 年 2 月被国际山茶花协会评为"国际杰出茶花园"。2022 年入选重庆市首批历史名园。由于前些年古茶树势有所衰退，自 2020 年 9 月份起公园通过深翻、打透气孔等措施对古茶树增施生物有机肥，以改善土壤质量，提高土壤有益菌含量；适时修剪，喷施酸性叶面肥，灌根增强根系活力等复壮处理。

　　Wanzhou Xishan Park was built in 1925. At that time, the Sichuan warlord Yang Sen settled in Wanzhou and the park was rebuilt based on the original Wanzhou Xishan Taoist Temple. The park is one of the earliest famous parks as well as one of the famous ancient camellia gardens nationwidely. There are a wild profusion of plants and trees in the park and the camellia trees were planted at the beginning. With nearly a hundred years of careful and patient cultivation, currently there are more than 110 Sichuan ancient camellia trees. In Feb. 2016, it was recognized as "International Camellia Garden of Excellence" by International Camellia Society. In 2022, the park was selected into the first batch of famous historic parks in Chongqing. In order to overcome the decreased growth potential of ancient camellia trees in the garden, the management office started to fertilize ancient camellia trees with bio-organic fertilizers to increase beneficial bacteria content in the soil, as well as digging deeply and drilling air holes around camellia trees since September 2020. Meanwhile, the management office also takes good care of camellia trees by properly prune the side branches, periodically spray acidic foliar fertilizer, as well as restore root vitality by root irrigation upon needs.

古1：紫金冠
C. japonica 'Zijinguan'

位于重庆市万州西山公园古茶园，N30°48′17″，E108°22′51″，海拔189.3 m。株高4.7 m，基围115.8 cm，冠幅6.6/7.1 m，树龄320年，花期1–4月，长势良好。管护单位重庆市万州区园林绿化管理中心。

Located in the Ancient Camellia Garden of Chongqing Wanzhou Xishan Park, with latitude 30°48′17″ N, longitude 108°22′51″ E, elevation 189.3 m, tree H 4.7 m, CG 115.8 cm, CD 6.6 m/7.1 m, 320 y, Fl. Jan. to Apr. Grows well. Managed and maintained by Chongqing Wanzhou District Landscaping Management Center.

古 2: 胭脂鳞
C. japonica 'Yanzhilin'

位于重庆市万州西山公园古茶园，N30°48′19″，E108°22′52″，海拔 189.7 m。株高 2.7 m，基围 37.7 cm，两分枝，最大分枝 27.6 cm，冠幅 3.1/3.3 m，树龄 110 年，花期 2–4 月，长势良好。管护单位重庆市万州区园林绿化管理中心。

Located in the Ancient Camellia Garden of Chongqing Wanzhou Xishan Park, with latitude 30°48′19″ N, longitude 108°22′52″ E, elevation 189.7 m, tree H 2.7 m, CG 37.7 cm, two branches (the biggest one 27.6 cm), CD 3.1 m/3.3 m, 110 y, Fl. Feb. to Apr. Grows well. Managed and maintained by Chongqing Wanzhou District Landscaping Management Center.

古 3：紫金冠
C. japonica 'Zijinguan'

位于重庆市万州西山公园古茶园，N30°48′19″，E108°22′52″，海拔 189.9 m。株高 3.8 m，基围 65.0 cm，三分枝，最大分枝 25.9 cm，冠幅 4.6/3.1 m，树龄 183 年，花期 1–4 月，基部有部分空腐，长势一般。管护单位重庆市万州区园林绿化管理中心。

Located in the Ancient Camellia Garden of Chongqing Wanzhou Xishan Park, with latitude 30°48′19″ N, longitude 108°22′52″ E, elevation 189.9 m, tree H 3.8 m, CG 65.0 cm, three branches (the biggest one 25.9 cm), CD 4.6 m/3.1 m, 183 y, Fl. Jan. to Apr. Some parts of trunk base are hollowly rotted. Grows in general. Managed and maintained by Chongqing Wanzhou District Landscaping Management Center.

古 4：洋红
C. japonica 'Yanghong'

位于重庆市万州西山公园五洲池，N30°48′16″，E108°22′49″，海拔 191.5 m。株高 3.2 m，基围 50.0 cm，冠幅 5.3/4.1 m，树龄 148 年，花期 1 月下旬至 3 月下旬，长势良好。管护单位重庆市万州区园林绿化管理中心。

Located in the Wuzhou Pool of Chongqing Wanzhou Xishan Park, with latitude 30°48′16″ N, longitude 108°22′49″ E, elevation 191.5 m, tree H 3.2 m, CG 50.0 cm, CD 5.3 m/4.1 m, 148 y, Fl. Jan. to Mar. Grows well. Managed and maintained by Chongqing Wanzhou District Landscaping Management Center.

古 5：紫金冠
C. japonica 'Zijinguan'

位于重庆市万州西山公园五洲池，N30°48′16″，E108°22′49″，海拔 190.1 m。株高 3.9 m，基围 95.8 cm，三分枝，最大分枝 32.3 cm，冠幅 4.1/4.3 m，树龄 222 年，花期 1–4 月，长势良好。管护单位重庆市万州区园林绿化管理中心。

Located in the Wuzhou Pool of Chongqing Wanzhou Xishan Park, with latitude 30°48′16″ N, longitude 108°22′49″ E, elevation 190.1 m, tree H 3.9 m, CG 95.8 cm, three branches (the biggest one 32.3 cm), CD 4.1 m/4.3 m, 222 y, Fl. Jan. to Apr. Grows well. Managed and maintained by Chongqing Wanzhou District Landscaping Management Center.

古6：白宝塔
C. japonica 'Baibaota'

位于重庆市万州西山公园五洲池，N30°48′15″，E108°22′49″，海拔 186.6 m。株高 3.1 m，基围 53.4 cm，冠幅 4.2/4.8 m，树龄 160 年，花期 2–4 月，长势良好。管护单位重庆市万州区园林绿化管理中心。

Located in the Wuzhou Pool of Chongqing Wanzhou Xishan Park, with latitude 30°48′15″ N, longitude 108°22′49″ E, elevation 186.6 m, tree H 3.1 m, CG 53.4 cm, CD 4.2 m/4.8 m, 160 y, Fl. Feb. to Apr. Grows well. Managed and maintained by Chongqing Wanzhou District Landscaping Management Center.

古 7: 洋红
C. japonica 'Yanghong'

位于重庆市万州西山公园古茶园，N30°48′17″，E108°22′52″，海拔186.5 m。株高4.1 m，基围57.1 cm，冠幅6.4/5.7 m，树龄126年，花期1月下旬至3月下旬。管护单位重庆市万州区园林绿化管理中心。

Located in the Ancient Camellia Garden of Chongqing Wanzhou Xishan Park, with latitude 30°48′17″ N, longitude 108°22′52″ E, elevation 186.5 m, tree H 4.1 m, CG 57.1 cm, CD 6.4 m/5.7 m, 126 y, Fl. Jan. to Mar. Managed and maintained by Chongqing Wanzhou District Landscaping Management Center.

古 8：紫金冠
C. japonica 'Zijinguan'

位于重庆市万州西山公园古茶园，N30°48′17″，E108°22′52″，海拔 186.6 m。株高 4.2 m，基围 88.8 cm，冠幅 5.4 m，树龄 205 年，花期 1–4 月。管护单位重庆市万州区园林绿化管理中心。

Located in the Ancient Camellia Garden of Chongqing Wanzhou Xishan Park, with latitude 30°48′17″ N, longitude 108°22′52″ E, elevation 186.6 m, tree H 4.2 m, CG 88.8 cm, CD 5.4 m, 205 y, Fl. Jan. to Apr. Managed and maintained by Chongqing Wanzhou District Landscaping Management Center.

古9：白六方
C. japonica 'Bailiufang'

位于万州西山公园古茶园，N30°48′15″，E108°22′49″，海拔 187.4 m。株高 3.7 m，基围 96.8 cm，冠幅 4.4 m，树龄 200 年，花期 2–4 月，长势良好。管护单位重庆市万州区园林绿化管理中心。

Located in the Ancient Camellia Garden of Wanzhou Xishan Park, with latitude 30°48′15″ N, longitude 108°22′49″ E, elevation 187.4 m, tree H 3.7 m, CG 96.8 cm, CD 4.4 m, 200 y, Fl. Feb. to Apr. Grows well. Managed and maintained by Chongqing Wanzhou District Landscaping Management Center.

古10：洋红
C. japonica 'Yanghong'

位于重庆市万州西山公园古茶园，N30°48′17″，E108°22′42″，海拔186.9 m。株高3.4 m，基围48.9 cm，冠幅3.4/4.7 m，树龄165年，花期1月卜旬至3月卜旬，长势良好。管护单位重庆市万州区园林绿化管理中心。

Located in the Ancient Camellia Garden of Chongqing Wanzhou Xishan Park, with latitude 30°48′17″ N, longitude 108°22′42″ E, elevation 186.9 m, tree H 3.4 m, CG 48.9 cm, CD 3.4 m/4.7 m, 165 y, Fl. Jan. to Mar. Grows well. Managed and maintained by Chongqing Wanzhou District Landscaping Management Center.

古11：金盘荔枝
C. japonica 'Jinpan Lizhi'

位于重庆市万州西山公园古茶园，N30°48′18″，E108°22′51″，海拔189.0 m。株高4.6 m，基围77.8 cm，冠幅8.5/5.2 m，树龄178年，花期2–3月，长势良好。管护单位重庆市万州区园林绿化管理中心。

Located in the Ancient Camellia Garden of Chongqing Wanzhou Xishan Park, with latitude 30°48′18″ N, longitude 108°22′51″ E, elevation 189.0 m, tree H 4.6 m, CG 77.8 cm, CD 8.5 m/5.2 m, 178 y, Fl. Feb. to Mar. Grows well. Managed and maintained by Chongqing Wanzhou District Landscaping Management Center.

古12：白宝塔
C. japonica 'Baibaota'

位于万州西山公园古茶园，N30°48′18″，E108°22′51″，海拔189.3 m。株高5.0 m，基围88.8 cm，冠幅8.1/5.7 m，树龄246年，花期2-4月，长势良好。管护单位重庆市万州区园林绿化管理中心。

Located in the Ancient Camellia Garden of Chongqing Wanzhou Xishan Park, with latitude 30°48′18″ N, longitude 108°22′51″ E, elevation 189.3 m, tree H 5.0 m, CG 88.8 cm, CD 8.1 m/5.7 m, 246 y, Fl. Feb. to Apr. Grows well. Managed and maintained by Chongqing Wanzhou District Landscaping Management Center.

古13：紫金冠
C. japonica 'Zijinguan'

位于重庆市万州西山公园古茶园，N30°48′17″，E108°22′51″，海拔189.4 m。株高4.7 m，基围82.1 cm，两分枝，最大分枝51.9 cm，冠幅6.2/8.5 m，树龄228年，花期1–4月，长势良好。管护单位重庆市万州区园林绿化管理中心。

Located in the Ancient Camellia Garden of Chongqing Wanzhou Xishan Park, with latitude 30°48′17″ N, longitude 108°22′51″ E, elevation 189.4 m, tree H 4.7 m, CG 82.1 cm, two branches (the biggest one 51.9 cm), CD 6.2 m/8.5 m, 228 y, Fl. Jan. to Apr. Grows well. Managed and maintained by Chongqing Wanzhou District Landscaping Management Center.

古14：胭脂鳞
C. japonica 'Yanzhilin'

位于重庆市万州西山公园古茶园，N30°48'18″，E108°22'51″，海拔191.5 m。株高4.3 m，基围56.5 cm，两分枝，最大分枝28.3 cm，冠幅4.3/4.7 m，树龄125年，花期2-4月，长势良好。管护单位重庆市万州区园林绿化管理中心。

Located in the Ancient Camellia Garden of Chongqing Wanzhou Xishan Park, with latitude 30°48'18″ N, longitude 108°22'51″ E, elevation 191.5 m, tree H 4.3 m, CG 56.5 cm, two branches (the biggest one 28.3 cm), CD 4.3 m/4.7 m, 125 y, Fl. Feb. to Apr. Grows well. Managed and maintained by Chongqing Wanzhou District Landscaping Management Center.

古 15：紫金冠
C. japonica 'Zijinguan'

位于重庆市万州西山公园古茶园，N30°48′17″，E108°22′51″，海拔 190.6 m。株高 6.4 m，基围 64.3 cm，冠幅 6.4 m，树龄 142 年，花期 1–4 月，长势良好。管护单位重庆市万州区园林绿化管理中心。

Located in the Ancient Camellia Garden of Chongqing Wanzhou Xishan Park, with latitude 30°48′17″ N, longitude 108°22′51″ E, elevation 190.6 m, tree H 6.4 m, CG 64.3 cm, CD 6.4 m, 142 y, Fl. Jan. to Apr. Grows well. Managed and maintained by Chongqing Wanzhou District Landscaping Management Center.

古16：七心红
C. japonica 'Qixinhong'

位于重庆市万州西山公园古茶园，N30°48'18"，E108°22'51"，海拔191.3 m。株高2.5 m，基围50.5 cm，冠幅3.5/3.1 m，树龄138年，花期2-4月，长势良好。管护单位重庆市万州区园林绿化管理中心。

Located in the Ancient Camellia Garden of Chongqing Wanzhou Xishan Park, with latitude 30°48'18" N, longitude 108°22'51" E, elevation 191.3 m, tree H 2.5 m, CG 50.5 cm, CD 3.5 m/3.1 m, 138 y, Fl. Feb. to Apr. Grows well. Managed and maintained by Chongqing Wanzhou District Landscaping Management Center.

古17：洋红
C. japonica 'Yanghong'

位于重庆市万州西山公园古茶园，N30°48′18″，E108°22′51″，海拔191.7 m。株高4.8 m，基围64.4 cm，冠幅5.4/8.7 m，树龄164年，花期1月下旬至3月下旬，长势良好。管护单位重庆市万州区园林绿化管理中心。

Located in the Ancient Camellia Garden of Chongqing Wanzhou Xishan Park, with latitude 30°48′18″ N, longitude 108°22′51″ E, elevation 191.7 m, tree H 4.8 m, CG 64.4 cm, CD 5.4 m/8.7 m, 164 y, Fl. Jan. to Mar. Grows well. Managed and maintained by Chongqing Wanzhou District Landscaping Management Center.

古 18：紫金冠
C. japonica 'Zijinguan'

位于重庆市万州西山公园古茶园，N30°48′17″，E108°22′51″，海拔 190.4 m。株高 6.0 m，基围 59.8 cm，两分枝，较大分枝基围 47.2 cm，冠幅 6.3/5.1 m，树龄 151 年，花期 1-4 月，长势良好。管护单位重庆市万州区园林绿化管理中心。

Located in the Ancient Camellia Garden of Chongqing Wanzhou Xishan Park, with latitude 30°48′17″ N, longitude 108°22′51″ E, elevation 190.4 m, tree H 6.0 m, CG 59.8 cm, two branches (the biggest one 47.2 cm), CD 6.3 m/5.1 m, 151 y, Fl. Jan. to Apr. Grows well. Managed and maintained by Chongqing Wanzhou District Landscaping Management Center.

古19：紫金冠
C. japonica 'Zijinguan'

位于重庆市万州西山公园古茶园，N30°48′18″，E108°22′50″，海拔191.7 m。株高6.0 m，基围86.6 cm，两分枝，最大分枝44.5 cm，冠幅6.3/5.1 m，树龄200年，花期1–4月，基部有部分空腐，长势良好。管护单位重庆市万州区园林绿化管理中心。

Located in the Ancient Camellia Garden of Chongqing Wanzhou Xishan Park, with latitude 30°48′18″ N, longitude 108°22′50″ E, elevation 191.7 m, tree H 6.0 m, CG 86.6 cm, two branches (the biggest one 44.5 cm), CD 6.3 m/5.1 m, 200 y, Fl. Jan. to Apr. Some parts of trunk base are hollowly rotted. Grows well. Managed and maintained by Chongqing Wanzhou District Landscaping Management Center.

古20: 胭脂鳞
C. japonica 'Yanzhilin'

位于重庆市万州西山公园古茶园，N30°48′17″，E108°22′52″，海拔 187.7 m。株高 3.1 m，基围 46.7 cm，冠幅 3.3/4.7 m，树龄 146 年，花期 2–4 月，基部枝干有部分空腐，长势良好。管护单位重庆市万州区园林绿化管理中心。

Located in the Ancient Camellia Garden of Chongqing Wanzhou Xishan Park, with latitude 30°48′17″ N, longitude 108°22′52″ E, elevation 187.7 m, tree H 3.1 m, CG 46.7 cm, CD 3.3 m/4.7 m, 146 y, Fl. Feb. to Apr. Some parts of trunk base are hollowly rotted. Grows well. Managed and maintained by Chongqing Wanzhou District Landscaping Management Center.

古 21: 紫金冠
C. japonica 'Zijinguan'

位于重庆市万州西山公园古茶园，N30°48′17″，E108°22′50″，海拔 191.6 m。株高 4.3 m，基围 81.0 cm，两分枝，最大分枝 38.6 cm，冠幅 5.3/4.1 m，树龄 186 年，花期 1–4 月，长势良好。管护单位重庆市万州区园林绿化管理中心。

Located in the Ancient Camellia Garden of Chongqing Wanzhou Xishan Park, with latitude 30°48′17″ N, longitude 108°22′50″ E, elevation 191.6 m, tree H 4.3 m, CG 81.0 cm, two branches (the biggest one 38.6 cm), CD 5.3 m/4.1 m, 186 y, Fl. Jan. to Apr. Grows well. Managed and maintained by Chongqing Wanzhou District Landscaping Management Center.

古 22: 胭脂鳞
C. japonica 'Yanzhilin'

位于万州西山公园古茶园，N30°48′18″，E108°22′52″，海拔188.5 m。株高4.2 m，基围52.1 cm，冠幅3.3/5.7 m，树龄143年，花期2–4月，长势良好。管护单位重庆市万州区园林绿化管理中心。

Located in the Ancient Camellia Garden of Chongqing Wanzhou Xishan Park, with latitude 30°48′18″ N, longitude 108°22′52″ E, elevation 188.5 m, tree H 4.2 m, CG 52.1 cm, CD 3.3 m/5.7 m, 143 y, Fl. Feb. to Apr. Grows well. Managed and maintained by Chongqing Wanzhou District Landscaping Management Center.

古 23：红佛鼎
C. japonica 'Hongfoding'

位于重庆市万州西山公园古茶园，N30°48′17″，E108°22′50″，海拔 191.4 m。株高 4.4 m，基围 62.8 cm，两分枝，最大分枝 27.0 cm，冠幅 5.3/4.1 m，偏冠，树龄 238 年，花期 2–4 月，长势良好。管护单位重庆市万州区园林绿化管理中心。

Located in the Ancient Camellia Garden of Chongqing Wanzhou Xishan Park, with latitude 30°48′17″ N, longitude 108°22′50″ E, elevation 191.4 m, tree H 4.4 m, CG 62.8 cm, two branches (the biggest one 27.0 cm), CD 5.3 m/4.1 m, 238 y, Fl. Feb. to Apr. Its canopy leans to one side. Grows well. Managed and maintained by Chongqing Wanzhou District Landscaping Management Center.

古 24：重庆红
C. japonica 'Chongqinghong'

位于重庆市万州西山公园古茶园，N30°48′18″，E108°22′51″，海拔 190.1 m。株高 3.8 m，基围 69.1 cm，冠幅 3.5/3.8 m，树龄 212 年，花期 3–5 月上旬，长势良好。管护单位重庆市万州区园林绿化管理中心。

Located in the Ancient Camellia Garden of Chongqing Wanzhou Xishan Park, with latitude 30°48′18″ N, longitude 108°22′51″ E, elevation 190.1 m, tree H 3.8 m, CG 69.1 cm, CD 3.5 m/3.8 m, 212 y, Fl. Mar. to May. Grows well. Managed and maintained by Chongqing Wanzhou District Landscaping Management Center.

古 25：紫金冠
C. japonica 'Zijinguan'

位于重庆市万州西山公园古茶园，N30°48'18″，E108°22'51″，海拔 190.4 m。株高 8.0 m，基围 69.8 cm，冠幅 4.6/6.1 m，树龄 156 年，花期 1-4 月，长势良好。管护单位重庆市万州区园林绿化管理中心。

Located in the Ancient Camellia Garden of Chongqing Wanzhou Xishan Park, with latitude 30°48'18″ N, longitude 108°22'51″ E, elevation 190.4 m, tree H 8.0 m, CG 69.8 cm, CD 4.6 m/6.1 m, 156 y, Fl. Jan. to Apr. Grows well. Managed and maintained by Chongqing Wanzhou District Landscaping Management Center.

古 26：重庆红
C. japonica 'Chongqinghong'

位于重庆市万州西山公园古茶园，N30°48′18″，E108°22′51″，海拔 189.1 m。株高 3.6 m，基围 81.2 cm，内分枝，最大分枝 68.5 cm，冠幅 4.8/4.2 m，树龄 206 年，花期 3-5 月上旬，长势良好。管护单位重庆市万州区园林绿化管理中心。

Located in the Ancient Camellia Garden of Chongqing Wanzhou Xishan Park, with latitude 30°48′18″ N, longitude 108°22′51″ E, elevation 189.1 m, tree H 3.6 m, CG 81.2 cm, two branches (the biggest one 68.5 cm), CD 4.8 m/4.2 m, 206 y, Fl. Mar. to May. Grows well. Managed and maintained by Chongqing Wanzhou District Landscaping Management Center.

古27：紫金冠
C. japonica 'Zijinguan'

位于重庆市万州西山公园古茶园，N30°48′18″，E108°22′50″，海拔192.6 m。株高4.0 m，基围48.3 cm，冠幅4.6/4.1 m，树龄161年，花期1–4月，长势良好。管护单位重庆市万州区园林绿化管理中心。

Located in the Ancient Camellia Garden of Chongqing Wanzhou Xishan Park, with latitude 30°48′18″ N, longitude 108°22′50″ E, elevation 192.6 m, tree H 4.0 m, CG 48.3 cm, CD 4.6 m/4.1 m, 161 y, Fl. Jan. to Apr. Grows well. Managed and maintained by Chongqing Wanzhou District Landscaping Management Center.

古 28：紫金冠
C. japonica 'Zijinguan'

位于重庆市万州西山公园古茶园，N30°48′18″，E108°22′51″，海拔 191.1 m。株高 5.0 m，基围 94.2 cm，两分枝，最大分枝 59.9 cm，冠幅 5.1/5.8 m，树龄 219 年，花期 1–4 月，长势良好。管护单位重庆市万州区园林绿化管理中心。

Located in the Ancient Camellia Garden of Chongqing Wanzhou Xishan Park, with latitude 30°48′18″ N, longitude 108°22′51″ E, elevation 191.1 m, tree H 5.0 m, CG 94.2 cm, two branches (the biggest one 59.9 cm), CD 5.1 m/5.8 m, 219 y, Fl. Jan. to Apr. Grows well. Managed and maintained by Chongqing Wanzhou District Landscaping Management Center.

古29：七心红
C. japonica 'Qixinhong'

位于重庆市万州西山公园古茶园，N30°48′18″，E108°22′51″，海拔 191.2 m。株高 3.6 m，基围 63.2 cm，冠幅 5.3/3.8 m，树龄 157 年，花期 2-4 月，长势良好。管护单位重庆市万州区园林绿化管理中心。

Located in the Ancient Camellia Garden of Chongqing Wanzhou Xishan Park, with latitude 30°48′18″ N, longitude 108°22′51″ E, elevation 191.2 m, tree H 3.6 m, CG 63.2 cm, CD 5.3 m/3.8 m, 157 y, Fl. Feb. to Apr. Grows well. Managed and maintained by Chongqing Wanzhou District Landscaping Management Center.

古 30：紫金冠
C. japonica 'Zijinguan'

位于重庆市万州西山公园古茶园，N30°48′18″，E108°22′51″，海拔190.8 m。株高4.4 m，基围68.5 cm，三分枝，最大分枝28.7 cm，冠幅3.3/2.1 m，树龄154年，花期1–4月，长势良好。管护单位重庆市万州区园林绿化管理中心。

Located in the Ancient Camellia Garden of Chongqing Wanzhou Xishan Park, with latitude 30°48′18″ N, longitude 108°22′51″ E, elevation 190.8 m, tree H 4.4 m, CG 68.5 cm, three branches (the biggest one 28.7 cm), CD 3.3 m/2.1 m, 154 y, Fl. Jan. to Apr. Grows well. Managed and maintained by Chongqing Wanzhou District Landscaping Management Center.

古31：紫金冠
C. japonica 'Zijinguan'

位于重庆市万州西山公园古茶园，N30°48′18″，E108°22′51″，海拔190.3 m。株高5.3 m，基围59.6 cm，冠幅5.3/4.1 m，树龄131年，花期1–4月，长势良好。管护单位重庆市万州区园林绿化管理中心。

Located in the Ancient Camellia Garden of Chongqing Wanzhou Xishan Park, with latitude 30°48′18″ N, longitude 108°22′51″ E, elevation 190.3 m, tree H 5.3 m, CG 59.6 cm, CD 5.3 m/4.1 m, 131 y, Fl. Jan. to Apr. Grows well. Managed and maintained by Chongqing Wanzhou District Landscaping Management Center.

古32：重庆红
C. japonica 'Chongqinghong'

位于重庆市万州西山公园古茶园，N30°48′18″，E108°22′51″，海拔 191.2 m。株高 5.3 m，基围 110.1 cm，冠幅 5.5/6.8 m，树龄 272 年，花期 3–5 月上旬，长势良好。管护单位重庆市万州区园林绿化管理中心。

Located in the Ancient Camellia Garden of Chongqing Wanzhou Xishan Park, with latitude 30°48′18″ N, longitude 108°22′51″ E, elevation 191.2 m, tree H 5.3 m, CG 110.1 cm, CD 5.5 m/6.8 m, 272 y, Fl. Mar. to May. Grows well. Managed and maintained by Chongqing Wanzhou District Landscaping Management Center.

古 33：重庆红
C. japonica 'Chongqinghong'

位于重庆市万州西山公园古茶园，N30°48′18″，E108°22′51″，海拔 189.2 m。株高 5.3 m，基围 110.4 cm，冠幅 5.1/6.7 m，树龄 259 年，花期 3–5 月上旬，长势良好。管护单位重庆市万州区园林绿化管理中心。

Located in the Ancient Camellia Garden of Chongqing Wanzhou Xishan Park, with latitude 30°48′18″ N, longitude 108°22′51″ E, elevation 189.2 m, tree H 5.3 m, CG 110.4 cm, CD 5.1 m/6.7 m, 259 y, Fl. Mar. to May. Grows well. Managed and maintained by Chongqing Wanzhou District Landscaping Management Center.

古34：紫金冠
C. japonica 'Zijinguan'

位于重庆市万州西山公园古茶园，N30°48′18″，E108°22′51″，海拔188.5 m。株高5.3 m，基围70.5 cm，冠幅5.3 m，树龄159年，花期1–4月，长势良好。管护单位重庆市万州区园林绿化管理中心。

Located in the Ancient Camellia Garden of Chongqing Wanzhou Xishan Park, with latitude 30°48′18″ N, longitude 108°22′51″ E, elevation 188.5 m, tree H 5.3 m, CG 70.5 cm, CD 5.3 m, 159 y, Fl. Jan. to Apr. Grows well. Managed and maintained by Chongqing Wanzhou District Landscaping Management Center.

古 35：白宝塔
C. japonica 'Baibaota'

位于重庆市万州西山公园古茶园，N30°48′18″，E108°22′52″，海拔 188.9 m。株高 4.4 m，基围 52.1 cm，冠幅 5.1/5.7 m，树龄 114 年，花期 2–4 月，长势良好。管护单位重庆市万州区园林绿化管理中心。

Located in the Ancient Camellia Garden of Chongqing Wanzhou Xishan Park, with latitude 30°48′18″ N, longitude 108°22′52″ E, elevation 188.9 m, tree H 4.4 m, CG 52.1 cm, CD 5.1 m/5.7 m, 114 y, Fl. Feb. to Apr. Grows well. Managed and maintained by Chongqing Wanzhou District Landscaping Management Center.

古36：白宝塔
C. japonica 'Baibaota'

位于重庆市万州西山公园古茶园，N30°48′18″，E108°22′52″，海拔 188.7 m。株高 3.3 m，基围 65.2 cm，冠幅 4.2/5.4 m，树龄 146 年，花期 2–4 月，长势良好。管护单位重庆市万州区园林绿化管理中心。

Located in the Ancient Camellia Garden of Chongqing Wanzhou Xishan Park, with latitude 30°48′18″ N, longitude 108°22′52″ E, elevation 188.7 m, tree H 3.3 m, CG 65.2 cm, CD 4.2 m/5.4 m, 146 y, Fl. Feb. to Apr. Grows well. Managed and maintained by Chongqing Wanzhou District Landscaping Management Center.

古 37：醉杨妃
C. japonica 'Zuiyangfei'

位于重庆市万州西山公园古茶园，N30°48′18″，E108°22′51″，海拔 192.7 m。株高 6.2 m，基围 55.4 cm，冠幅 6.8/5.9 m，树龄 130 年，花期 2–4 月，长势良好。管护单位重庆市万州区园林绿化管理中心。

Located in the Ancient Camellia Garden of Chongqing Wanzhou Xishan Park, with latitude 30°48′18″ N, longitude 108°22′51″ E, elevation 192.7 m, tree H 6.2 m, CG 55.4 cm, CD 6.8 m/5.9 m, 130 y, Fl. Feb. to Apr. Grows well. Managed and maintained by Chongqing Wanzhou District Landscaping Management Center.

古 38：胭脂鳞
C. japonica 'Yanzhilin'

位于重庆市万州西山公园古茶园，N30°48′18″，E108°22′51″，海拔 192.1 m。株高 3.5 m，基围 62.5 cm，冠幅 4.3/4.7 m，树龄 133 年，花期 2-4 月，长势良好。管护单位重庆市万州区园林绿化管理中心。

Located in the Ancient Camellia Garden of Chongqing Wanzhou Xishan Park, with latitude 30°48′18″ N, longitude 108°22′51″ E, elevation 192.1 m, tree H 3.5 m, CG 62.5 cm, CD 4.3 m/4.7 m, 133 y, Fl. Feb. to Apr. Grows well. Managed and maintained by Chongqing Wanzhou District Landscaping Management Center.

古 39：紫金冠
C. japonica 'Zijinguan'

位于重庆市万州西山公园花卉园，N30°48′21″，E108°22′52″，海拔 195.2 m。株高 4.4 m，基围 48.2 cm，冠幅 3.3/4.1 m，树龄 100 年，花期 2-4 月，长势良好。管护单位重庆市万州区园林绿化管理中心。

Located in the Flower Garden of Chongqing Wanzhou Xishan Park, with latitude 30°48′21″ N, longitude 108°22′52″ E, elevation 195.2 m, tree H 4.4 m, CG 48.2 cm, CD 3.3 m/4.1 m, 100 y, Fl. Feb. to Apr. Grows well. Managed and maintained by Chongqing Wanzhou District Landscaping Management Center.

古40: 洋红
C. japonica 'Yanghong'

位于重庆市万州西山公园花卉园,N30°48′21″,E108°22′52″,海拔194.9 m。株高4.2m,基围56.2 cm,三分枝,最大分枝26.7 cm,冠幅4.4/4.7 m,树龄143年,花期1月下旬至3月下旬,长势良好。管护单位重庆市万州区园林绿化管理中心。

Located in the Flower Garden of Chongqing Wanzhou Xishan Park, with latitude 30°48′21″ N, longitude 108°22′52″ E, elevation 194.9 m, tree H 4.2 m, CG 56.2 cm, three branches (the biggest one 26.7 cm), CD 4.4 m/4.7 m, 143 y, Fl. Jan. to Mar. Grows well. Managed and maintained by Chongqing Wanzhou District Landscaping Management Center.

古 41：紫金冠
C. japonica 'Zijinguan'

位于重庆市万州西山公园花卉园，N30°48′22″，E108°22′53″，海拔 195.0 m。株高 3.2 m，基围 64.5 cm，冠幅 4.1/5.8 m，树龄 171 年，花期 1–4 月，长势良好。管护单位重庆市万州区园林绿化管理中心。

Located in the Flower Garden of Chongqing Wanzhou Xishan Park, with latitude 30°48′22″ N, longitude 108°22′53″ E, elevation 195.0 m, tree H 3.2 m, CG 64.5 cm, CD 4.1 m/5.8 m, 171 y, Fl. Jan. to Apr. Grows well. Managed and maintained by Chongqing Wanzhou District Landscaping Management Center.

古42：金盘荔枝
C. japonica 'Jinpan Lizhi'

位于重庆市万州西山公园花卉园，N30°48'21"，E108°22'52"，海拔195.5 m。株高3.5 m，基围47.8 cm，三分枝，最大分枝27.0 cm，冠幅3.5/3.2 m，树龄100年，花期2-3月，长势良好。管护单位重庆市万州区园林绿化管理中心。

Located in the Flower Garden of Chongqing Wanzhou Xishan Park, with latitude 30°48'21" N, longitude 108°22'52" E, elevation 195.5 m, tree H 3.5 m, CG 47.8 cm, three branches (the biggest one 27.0 cm), CD 3.5 m/3.2 m, 100 y, Fl. Feb. to Mar. Grows well. Managed and maintained by Chongqing Wanzhou District Landscaping Management Center.

古43：七心红
C. japonica 'Qixinhong'

位于重庆市万州西山公园花卉园，N30°48′22″，E108°22′53″，海拔189.5 m。株高2.0 m，基围20.1 cm，八分枝，最大分枝21.0 cm，冠幅4.1/3.3 m，树龄122年，花期2–4月，长势良好。管护单位重庆市万州区园林绿化管理中心。

Located in the Flower Garden of Chongqing Wanzhou Xishan Park, with latitude 30°48′22″ N, longitude 108°22′53″ E, elevation 189.5 m, tree H 2.0 m, CG 20.1 cm, eight branches (the biggest one 21.0 cm), CD 4.1 m/3.3 m, 122 y, Fl. Feb. to Apr. Grows well. Managed and maintained by Chongqing Wanzhou District Landscaping Management Center.

古 44：紫金冠
C. japonica 'Zijinguan'

位于重庆市万州西山公园古茶园，N30°48′18″，E108°22′52″，海拔 189.4 m。株高 3.7 m，基围 63.2 cm，三分枝，最大分枝 31.2 cm，冠幅 2.1/2.8 m，偏冠，树龄 140 年，花期 1–4 月，长势良好。管护单位重庆市万州区园林绿化管理中心。

Located in the Ancient Camellia Garden of Chongqing Wanzhou Xishan Park, with latitude 30°48′18″ N, longitude 108°22′52″ E, elevation 189.4 m, tree H 3.7 m, CG 63.2 cm, three branches (the biggest one 31.2 cm), CD 2.1 m/2.8 m, 140 y, Fl. Jan. to Apr. Its canopy leans to one side. Grows well. Managed and maintained by Chongqing Wanzhou District Landscaping Management Center.

古 45：七心红
C. japonica 'Qixinhong'

位于重庆市万州西山公园钟楼，N30°48′21″，E108°22′58″，海拔 171.1 m。株高 3.0 m，基围 54.2 cm，冠幅 3.5/4.1 m，偏冠，树龄 162 年，花期 2–4 月，长势良好。管护单位重庆市万州区园林绿化管理中心。

Located in the Bell Tower of Chongqing Wanzhou Xishan Park, with latitude 30°48′21″ N, longitude 108°22′58″ E, elevation 171.1 m, tree H 3.0 m, CG 54.2 cm, CD 3.5 m/4.1 m, 162 y, Fl. Feb. to Apr. Its canopy leans to one side. Grows well. Managed and maintained by Chongqing Wanzhou District Landscaping Management Center.

古46：七心红
C. japonica 'Qixinhong'

位于重庆市万州西山公园主路旁，N30°48′20″，E108°22′55″，海拔181.4 m。株高3.2 m，基围62.9 cm，冠幅3.1 m，树龄171年，花期2–4月，长势良好。管护单位重庆市万州区园林绿化管理中心。

Located at the main roadside of Chongqing Wanzhou Xishan Park, with latitude 30°48′20″ N, longitude 108°22′55″ E, elevation 181.4 m, tree H 3.2 m, CG 62.9 cm, CD 3.1 m, 171 y, Fl. Feb. to Apr. Grows well. Managed and maintained by Chongqing Wanzhou District Landscaping Management Center.

古47: 白宝塔
C. japonica 'Baibaota'

位于重庆市万州西山公园主路旁，N30°48′21″，E108°22′56″，海拔181.1 m。株高2.8 m，基围63.5 cm，冠幅4.1/2.3 m，树龄174年，花期2–4月，长势良好。管护单位重庆市万州区园林绿化管理中心。

Located at the main roadside of Chongqing Wanzhou Xishan Park, with latitude 30°48′21″ N, longitude 108°22′56″ E, elevation 181.1 m, tree H 2.8 m, CG 63.5 cm, CD 4.1 m/2.3 m, 174 y, Fl. Feb. to Apr. Grows well. Managed and maintained by Chongqing Wanzhou District Landscaping Management Center.

古48：紫金冠
C. japonica 'Zijinguan'

位于万州西山公园钟楼，N30°48′21″，E108°22′57″，海拔172.3 m。株高4.5 m，基围63.8 cm，两分枝，最大分枝34.3 cm，冠幅3.8/4.2 m，树龄168年，花期1–4月，长势良好。管护单位重庆市万州区园林绿化管理中心。

Located in the Bell Tower of Chongqing Wanzhou Xishan Park, with latitude 30°48′21″ N, longitude 108°22′57″ E, elevation 172.3 m, tree H 4.5 m, CG 63.8 cm, two branches (the biggest one 34.3 cm), CD 3.8 m/4.2 m, 168 y, Fl. Jan. to Apr. Grows well. Managed and maintained by Chongqing Wanzhou District Landscaping Management Center.

古49：七心白
C. japonica 'Qixinbai'

位于重庆市万州西山公园五洲池，N30°48′14″，E108°22′50″，海拔187.3 m。株高3.1 m，基围50.3 cm，冠幅5.3/3.7 m，树龄168年，花期2-4月，长势良好。管护单位重庆市万州区园林绿化管理中心。

Located in the Wuzhou Pool of Chongqing Wanzhou Xishan Park, with latitude 30°48′14″ N, longitude 108°22′50″ E, elevation 187.3 m, tree H 3.1 m, CG 50.3 cm, CD 5.3 m/3.7 m, 168 y, Fl. Feb. to Apr. Grows well. Managed and maintained by Chongqing Wanzhou District Landscaping Management Center.

古50：闽鄂山茶
C. grijsii Hance

位于重庆市万州西山公园花卉园，N30°48′21″，E108°22′52″，海拔 194.6 m。株高 8.0 m，基围 71.2 cm，冠幅 7.3/5.7 m，树龄 216 年，花期 12 月至翌年 1 月，长势良好。管护单位重庆市万州区园林绿化管理中心。

Located in the Flower Garden of Chongqing Wanzhou Xishan Park, with latitude 30°48′21″ N, longitude 108°22′52″ E, elevation 194.6 m, tree H 8.0 m, CG 71.2 cm, CD 7.3 m/5.7 m, 216 y, Fl. Dec. to Jan. Grows well. Managed and maintained by Chongqing Wanzhou District Landscaping Management Center.

万州区
分水镇枣园

INTRODUCTION TO ZAOYUAN VILLAGE, FENSHUI TOWN, WANZHOU DISTRICT

古1：花洋红
C. japonica 'Hua Yanghong'

位于重庆市万州区分水镇枣园村民秦家，N30°43′03″，E108°08′27″，海拔624.0 m。株高5.5 m，基围95.0 cm，冠幅7.3 m，树龄139年，花期1–4月，长势良好。据该树主人介绍，20世纪20年代其曾祖父在万州西山公园工作时搬回家栽种。管护单位重庆市万州区绿化委员会。

Located in the villager Qin's house at Zaoyuan Village in Fenshui Town, Wanzhou District, Chongqing, with latitude 30°43′03″ N, longitude 108°08′27″ E, elevation 624.0 m, tree H 5.5 m, CG 95.0 cm, CD 7.3 m, 139 y, Fl. Jan. to Apr. Grows well. According to its owner, it was planted by his great-grandfather when working in Xishan Park in 1920s. Managed and maintained by Chongqing Wanzhou District Landscaping Committee.

万州区
恒合乡黄桐寨

INTRODUCTION TO HUANGTONG VILLAGE, HENGHE TOWN, WANZHOU DISTRICT

古1：西南红山茶
C. pitardii Coh. St. var. *Pitardii* Sealy

位于重庆市万州区恒合乡黄桐寨，N30°36′40″，E108°45′45″，海拔1180.1 m。株高5.3 m，基围57.0 cm，冠幅4.3/4.7 m，树龄128年。据当地人说该茶花在此处已历经五代人，五代人之前就有这么大，这些年因环境受限长势缓慢。管护单位重庆市万州区龙驹国有林场。

Located in Huangtong Village of Henghe Town, Wanzhou District, Chongqing, with latitude 30°36′40″ N, longitude 108°45′45″ E, elevation 1180.1 m, tree H 5.3 m, CG 57.0 cm, CD 4.3 m/4.7 m, 128 y. According to local people, it lives here at least five generations. Due to environmental constraints, it grows slowly. Managed and maintained by Chongqing Wanzhou District Longju State-owned Forest Farm.

西南大学

INTRODUCTION TO SOUTHWEST UNIVERSITY

西南大学溯源于1906年建立的川东师范学院，几经传承演变，1946年建立西南农学院，1985年更名为西南农业大学，2005年与西南师范大学合并为西南大学。校园内现有川山茶大树古树26株，集中栽植于原西南农业大学大门入口至行政大楼前的广场上。据记载这批古树是建校初期从原江北县静观镇和江北花朝门苗圃等地移入，后因校园建设重新移栽，现普遍长势较弱，需加强保护。

Southwest University can be traced back to East Sichuan Normal University established in 1906. Southwest Agricultural College, established in 1946 through inheritance and evolution of East Sichuan Normal University, was later renamed to Southwest Agricultural University in 1985, and then merged with Southwest Normal University as Southwest University in 2005. There are 26 Sichuan ancient and big camellia trees, and they are all planted on the square from the entrance of the former Southwest Agricultural University to the front of the administrative building. According to literature records, those ancient trees were transplanted from Jingguan Town, Jiangbei Huachaomen Nursery and other places when the university was initially constructed. Due to campus reconstruction, those camellia trees were replanted and lead to decreased growth potential. Thus, those camellia trees need better protection and caretaking.

古1：白洋片
C. japonica 'Baiyangpian'

位于西南大学法学院前，N29°49′02″，E106°24′52″，海拔 204.7 m。株高 2.5 m，基围 46.5 cm，冠幅 2.3 m，树龄 101 年，花期 1–4 月。由于校园环境改造，填埋土壤太深，长势差。管护单位西南大学。

Located in front of the Law School of Southwest University, with latitude 29°49′02″ N, longitude 106°24′52″ E, elevation 204.7 m, tree H 2.5 m, CG 46.5 cm, CD 2.3 m, 101 y, Fl. Jan. to Apr. Due to the renovation of the campus, filling soil around it is too deep, thus it grows weakly. Managed and maintained by Southwest University.

古2: 白宝塔
C. japonica 'Baibaota'

位于西南大学法学院前，N29°49′01″，E106°24′52″，海拔 205.6 m。株高 2.6 m，基围 57.0 cm，三分枝，基围 25.0/23.0/36.0 cm，冠幅 3.3 m，树龄 125 年，花期 2–4 月，长势一般。管护单位西南大学。

Located in front of the Law School of Southwest University, with latitude 29°49′01″ N, longitude 106°24′52″ E, elevation 205.6 m, tree H 2.6 m, CG 57.0 cm, three branches (25.0 cm/23.0 cm/36.0 cm), CD 3.3 m, 125 y, Fl. Feb. to Apr. Grows in general. Managed and maintained by Southwest University.

古 3：醉杨妃
C. japonica 'Zuiyangfei'

位于西南大学法学院前，N29°49′01″，E106°24′52″，海拔 206.1 m。株高 3.1 m，两分枝，基围 41.0 cm，冠幅 2.1/2.4 m，树龄 103 年，花期 2-4 月。由于校园环境改造，填埋土壤太深，枝干上可见明显孔洞和截枝痕迹，树皮皴裂，长势一般。管护单位西南大学。

Located in front of the Law School of Southwest University, with latitude 29°49′01″ N, longitude 106°24′52″ E, elevation 206.1 m, tree H 3.1 m, two branches, CG 41.0 cm, CD 2.1 m/2.4 m, 103 y, Fl. Feb. to Apr. Due to reconstruction of campus, filling soil around it is too deep. There are obvious holes and truncation marks on trunk. Grows in general. Managed and maintained by Southwest University.

古 4: 小红莲
C. japonica 'Xiaohonglian'

  位于西南大学法学院前，N29°49′01″，E106°24′52″，海拔 206.0 m。株高 3.3 m，基围 47.0 cm，冠幅 1.5/2.0 m，树龄 101 年，花期 3–5 月。由于校园环境改造，填埋土壤太深，枝叶稀疏，长势差。管护单位西南大学。

  Located in front of the Law School of Southwest University, with latitude 29°49′01″ N, longitude 106°24′52″ E, elevation 206.0 m, tree H 3.3 m, CG 47.0 cm, CD 1.5 m/2.0 m, 101 y, Fl. Mar. to May. Due to reconstruction of campus, filling soil around it is too deep. It grows weakly with sparse foliage. Managed and maintained by Southwest University.

古 5：白洋片
C. japonica 'Baiyangpian'

位于西南大学法学院前，N29°49′01″，E106°24′52″，海拔 205.9 m。株高 2.4 m，基围 45.0 cm，冠幅 2.6 m，树龄 101 年，花期 1–4 月，长势良好。管护单位西南大学。

Located in front of the Law School of Southwest University, with latitude 29°49′01″ N, longitude 106°24′52″ E, elevation 205.9 m, tree H 2.4 m, CG 45.0 cm, CD 2.6 m, 101 y, Fl. Jan. to Apr. Grows well. Managed and maintained by Southwest University.

古 6：白洋片
C. japonica 'Baiyangpian'

位于西南大学法学院前，N29°49'01"，E106°24'52"，海拔 238.4 m。株高 2.6 m，基围 57.0 cm，冠幅 2.5/2.2 m，树龄 125 年，花期 1–4 月。树干有孔洞，基部树皮皱裂。由于校园环境改造，填埋土壤太深，长势一般。管护单位西南大学。

Located in front of the Law School of Southwest University, with latitude 29°49'01" N, longitude 106°24'52" E, elevation 238.4 m, tree H 2.6 m, CG 57.0 cm, CD 2.5 m/2.2 m, 125 y, Fl. Jan. to Apr. There are obvious holes on trunk, and cracked bark at trunk base. Due to reconstruction of campus, filling soil around it is too deep. Grows in general. Managed and maintained by Southwest University.

古 7：白洋片
C. japonica 'Baiyangpian'

位于西南大学蚕桑纺织与生物研究所门前，N29°49′00″，E106°24′52″，海拔 205.3 m。株高 3.1 m，基围 45.0 cm，冠幅 3.6/2.9 m，树龄 101 年，花期 1–4 月。长势良好。管护单位西南大学。

Located in front of the Sericulture Textile and Biological Research Institute of Southwest University, with latitude 29°49′00″ N, longitude 106°24′52″ E, elevation 205.3 m, tree H 3.1 m, CG 45.0 cm, CD 3.6 m/2.9 m, 101 y, Fl. Jan. to Apr. Grows well. Managed and maintained by Southwest University.

古8：紫金冠
C. japonica 'Zijinguan'

位于西南大学蚕桑纺织与生物研究所门前，N29°49′00″，E106°24′52″，海拔205.5 m。株高2.6 m，五分枝，基围48.0 cm，冠幅2.6 m，树龄103年，花期1-4月。由于校园环境改造，填埋土壤太深，可见多数枯死枝，长势一般。管护单位西南大学。

Located in front of the Sericulture Textile and Biological Research Institute of Southwest University, with latitude 29°49′00 N, longitude 106°24′52″ E, elevation 205.5 m, tree H 2.6 m, five branches, CG 48.0 cm, CD 2.6 m, 103 y, Fl. Jan. to Apr. Due to reconstruction of campus, filling soil around it is too deep. Multiple dead twigs can be observed. Grows in general. Managed and maintained by Southwest University.

古9：紫金冠
C. japonica 'Zijinguan'

位于西南大学蚕桑纺织与生物研究所门前，N29°49′00″，E106°24′52″，海拔 204.8 m。株高 2.8 m，三分枝，基围 48.0 cm，冠幅 3.1 m，树龄 103 年，花期 1–4 月。管护单位西南大学。

Located in front of the Sericulture Textile and Biological Research Institute of Southwest University, with latitude 29°49′00″ N, longitude 106°24′52″ E, elevation 204.8 m, tree H 2.8 m, three branches, CG 48.0 cm, CD 3.1 m, 103 y, Fl. Jan. to Apr. Managed and maintained by Southwest University.

古10：白洋片
C. japonica 'Baiyangpian'

位于西南大学蚕桑纺织与生物研究所门前，N29°49′00″，E106°24′52″，海拔205.6 m。株高3.3 m，围径65.0 cm，冠幅3.1/4.0 m，树龄144年，花期1–4月，长势一般。管护单位西南大学。

Located in front of the Sericulture Textile and Biological Research Institute of Southwest University, with latitude 29°49′00″ N, longitude 106°24′52″ E, elevation 205.6 m, tree H 3.3 m, CG 65.0 cm, CD 3.1 m/4.0 m, 144 y, Fl. Jan. to Apr. Grows in general. Managed and maintained by Southwest University.

古 11：七心白
C. japonica 'Qixinbai'

位于西南大学蚕桑纺织与生物研究所门前，N29°48′59″，E106°24′52″，海拔 205.8 m。株高 3.5 m，基围 72.5 cm，冠幅 3.9/4.8 m，树龄 161 年，花期 2-4 月。有明显枯死枝和截枝痕迹，长势一般。管护单位西南大学。

Located in front of the Sericulture Textile and Biological Research Institute of Southwest University, with latitude 29°48′59″ N, longitude 106°24′52″ E, elevation 205.8 m, tree H 3.5 m, CG 72.5 cm, CD 3.9 m/4.8 m, 161 y, Fl. Feb. to Apr. There are obvious dead twigs and truncation marks. Grows in general. Managed and maintained by Southwest University.

古12：洋红
C. japonica 'Yanghong'

位于西南大学蚕桑纺织与生物研究所门前，N29°48′59″，E106°24′52″，海拔207.4 m。株高4.3 m，基围69.0 cm，冠幅5.0/3.9 m，树龄153年，花期2–4月，长势一般。管护单位西南大学。

Located in front of the Sericulture Textile and Biological Research Institute of Southwest University, with latitude 29°48′59″ N, longitude 106°24′52″ E, elevation 207.4 m, tree H 4.3 m, CG 69.0 cm, CD 5.0 m/3.9 m, 153 y, Fl. Feb. to Apr. Grows in general. Managed and maintained by Southwest University.

古 13：小红莲
C. japonica 'Xiaohonglian'

位于西南大学蚕桑纺织与生物研究所门前，N29°49'00″，E106°24'52″，海拔 206.5 m。株高 2.6 m，基围 41.0 cm，冠幅 2.9/2.6 m，树龄 105 年，花期 3–5 月，长势良好。管护单位西南大学。

Located in front of the Sericulture Textile and Biological Research Institute of Southwest University, with latitude 29°49'00″ N, longitude 106°24'52″ E, elevation 206.5 m, tree H 2.6 m, CG 41.0 cm, CD 2.9 m/2.6 m, 105 y, Fl. Mar. to May. Grows well. Managed and maintained by Southwest University.

古14：七心红
C. japonica 'Qixinhong'

位于西南大学蚕桑纺织与生物研究所门前，N29°48′59″，E106°24′52″，海拔206.8 m。株高2.3 m，基围50.0 cm，冠幅1.5/1.8 m，树龄108年，花期1–4月。主干已死，剩下侧枝，长势一般。管护单位西南大学。

Located in front of the Sericulture Textile and Biological Research Institute of Southwest University, with latitude 29°48′59″ N, longitude 106°24′52″ E, elevation 206.8 m, tree H 2.3 m, CG 50.0 cm, CD 1.5 m/1.8 m, 108 y, Fl. Jan. to Apr. The main trunk is dead, and several lateral branches still grow. Managed and maintained by Southwest University.

古15：金顶大红
C. japonica 'Jinding Dahong'

位于西南大学行政楼前，N29°49′01″，E106°24′49″，海拔211.6 m。株高3.1 m，两分枝，基围50.0 cm，冠幅4.3 m，树龄108年，花期1–4月。由于校园环境改造，填埋土壤太深，急需采取保护措施，现长势一般。管护单位西南大学。

Located in front of the Administration Building of Southwest University, with latitude 29°49′01″ N, longitude 106°24′49″ E, elevation 211.6 m, tree H 3.1 m, two branches, CG 50.0 cm, CD 4.3 m, 108 y, Fl. Jan. to Apr. Due to the renovation of the campus, the filling soil around is too deep, and lead to weak growth. It is urgent to take protection measures. Grows in general. Managed and maintained by Southwest University.

沙坪公园

INTRODUCTION TO
SHAPING PARK

沙坪公园为20世纪30年代开明绅士杨若愚的私家花园"愚庐"，现有茶花园是在"愚庐"的基础上多次扩建而成，于1957年4月正式开园。园内收集了多种名贵花木，现有川山茶古树16棵。2022年入选重庆市首批历史名园。

Shaping Park can be traced back to 1930s, is originally the private garden "Yulu" of an enlightened gentry, Yang Ruoyu. The current camellia garden, officially opened to public in April 1957, was built with several extensions on the basis of "Yulu". A variety of precious flowers and trees are preserved in it, including 16 Sichuan ancient camellia trees. In 2022, the park was selected into the first batch of famous historic parks in Chongqing.

古1：白洋片
C. japonica 'Baiyangpian'

位于重庆市沙坪公园茶花园，N29°33′12″，E106°27′09″，海拔242.9 m。株高2.9 m，基围36.0 cm，冠幅4.3/3.5 m，树龄162年，花期1–3月，长势一般。管护单位重庆市沙坪公园管理处。

Located in the Camellia Garden of Chongqing Shaping Park, with latitude 29°33′12″ N, longitude 106°27′09″ E, elevation 242.9 m, tree H 2.9 m, CG 36.0 cm, CD 4.3 m/3.5 m, 162 y, Fl. Jan. to Mar. Grows in general. Managed and maintained by Chongqing Shaping Park Administrative Office.

古 2: 白洋片
C. japonica 'Baiyangpian'

位于重庆市沙坪公园茶花园，N29°33′12″，E106°27′09″，海拔 242.4 m。株高 2.8 m，基围 40.0 cm，冠幅 3.4/2.1 m，树龄 175 年，花期 1–3 月，长势良好。管护单位重庆市沙坪公园管理处。

Located in the Camellia Garden of Chongqing Shaping Park, with latitude 29°33′12″ N, longitude 106°27′09″ E, elevation 242.4 m, tree H 2.8 m, CG 40.0 cm, CD 3.4 m/2.1 m, 175 y, Fl. Jan. to Mar. Grows well. Managed and maintained by Chongqing Shaping Park Administrative Office.

古3：白洋片
C. japonica 'Baiyangpian'

位于重庆市沙坪公园茶花园，N29°33'12"，E106°27'07"，海拔240.1 m。株高3.4 m，树干三分枝，基围45.0 cm，冠幅3.6/3.2 m，树龄190年，花期1–3月，枝干树皮有裂纹，长势良好。管护单位重庆市沙坪公园管理处。

Located in the Camellia Garden of Chongqing Shaping Park, with latitude 29°33'12" N, longitude 106°27'07" E, elevation 240.1 m, tree H 3.4 m, three branches, CG 45.0 cm, CD 3.6 m/3.2 m, 190 y, Fl. Jan. to Mar. Cracks on bark. Grows well. Managed and maintained by Chongqing Shaping Park Administrative Office.

古4：白洋片
C. japonica 'Baiyangpian'

位于重庆市沙坪公园茶花园，N29°33'12″，E106°27'09″，海拔 240.8 m，株高 4.2 m，树十二分枝，基围 60.0 cm，冠幅 3.8 m，树龄 237 年，花期 1-3 月，长势良好。管护单位重庆市沙坪公园管理处。

Located in the Camellia Garden of Chongqing Shaping Park, with latitude 29°33'12″ N, longitude 106°27'09″ E, elevation 240.8 m, tree H 4.2 m, three branches, CG 60.0 cm, CD 3.8 m, 237 y, Fl. Jan. to Mar. Grows well. Managed and maintained by Chongqing Shaping Park Administrative Office.

古 5：七心红
C. japonica 'Qixinhong'

位于重庆市沙坪公园茶花园，N29°33′11″，E106°27′09″，海拔 236.5 m。株高 2.1 m，基围 40.0 cm，冠幅 1.8/2.9 m，树龄 175 年，花期 1–4 月，分枝处呈瘤状凸起，长势一般。管护单位重庆市沙坪公园管理处。

Located in the Camellia Garden of Chongqing Shaping Park, with latitude 29°33′11″ N, longitude 106°27′09″ E, elevation 236.5 m, tree H 2.1 m, CG 40.0 cm, CD 1.8 m/2.9 m, 175 y, Fl. Jan. to Apr. There are tumor-like protrusions on branches. Grows in general. Managed and maintained by Chongqing Shaping Park Administrative Office.

古6：七心红
C. japonica 'Qixinhong'

位于重庆市沙坪公园茶花园，N29°33′12″，E106°27′08″，海拔241.8 m。株高3.5 m，树十六分枝，最大三枝基围50.0/28.0/19.0 cm，冠幅4.5/3.9 m，树龄206年，花期1–4月，树皮有不同程度脱落和开裂，长势差。管护单位重庆市沙坪公园管理处。

Located in the Camellia Garden of Chongqing Shaping Park, with latitude 29°33′12″ N, longitude 106°27′08″ E, elevation 241.8 m, tree H 3.5 m, six branches (the biggest three 50.0 cm/28.0 cm/19.0 cm respectively), CD 4.5 m/3.9 m, 206 y, Fl. Jan. to Apr. There are various degrees of peeling and splits on bark and trunk. Grows weakly. Managed and maintained by Chongqing Shaping Park Administrative Office.

古7：紫金冠
C. japonica 'Zijinguan'

位于重庆市沙坪公园茶花园入口处，N29°33′12″，E106°27′06″，海拔242.2 m。株高6.0 m，基围55.0 cm，冠幅6.0/4.7 m，树龄221年，花期2–4月，长势一般。管护单位重庆市沙坪公园管理处。

Located at the entrance of the Camellia Garden of Chongqing Shaping Park, with latitude 29°33′12″ N, longitude 106°27′06″ E, elevation 242.2 m, tree H 6.0 m, CG 55.0 cm, CD 6.0 m/4.7 m, 221 y, Fl. Feb. to Apr. Grows in general. Managed and maintained by Chongqing Shaping Park Administrative Office.

古 8：白洋片
C. japonica 'Baiyangpian'

位于重庆市沙坪公园茶花园入口处，N29°33′12″，E106°27′06″，海拔 242.9 m。株高 4.8 m，基围 52.0 cm，冠幅 4.2 m，树龄 212 年，花期 2–4 月，长势良好。管护单位重庆市沙坪公园管理处。

Located at the entrance of the Camellia Garden of Chongqing Shaping Park, with latitude 29°33′12″ N, longitude 106°27′06″ E, elevation 242.9 m, tree H 4.8 m, CG 52.0 cm, CD 4.2 m, 212 y, Fl. Feb. to Apr. Grows well. Managed and maintained by Chongqing Shaping Park Administrative Office.

古9：紫金冠
C. japonica 'Zijinguan'

位于重庆市沙坪公园茶花园入口处，N29°33′12″，E106°27′06″，海拔243.5 m。株高4.0 m，基围46.0 cm，冠幅3.4 m，树龄193年，花期2-4月。受一旁大树影响，树冠偏向一侧，长势良好。管护单位重庆市沙坪公园管理处。

Located at the entrance of the Camellia Garden of Chongqing Shaping Park, with latitude 29°33′12″ N, longitude 106°27′06″ E, elevation 243.5 m, tree H 4.0 m, CG 46.0 cm, CD 3.4 m, 193 y, Fl. Feb. to Apr. Affected by the nearby giant tree, its canopy leans to one side. Grows well. Managed and maintained by Chongqing Shaping Park Administrative Office.

古10: 白洋片
C. japonica 'Baiyangpian'

位于重庆市沙坪公园茶花园入口处，N29°33′12″，E106°27′06″，海拔246.4 m。株高4.5 m，基围60.0 cm，冠幅3.1/3.4 m，树龄237年，花期1–4月。枝干树皮有裂纹，受一旁大树影响，树冠偏向一侧，长势良好。管护单位重庆市沙坪公园管理处。

Located at the entrance of the Camellia Garden of Chongqing Shaping Park, with latitude 29°33′12″ N, longitude 106°27′06″ E, elevation 246.4 m, tree H 4.5 m, CG 60.0 cm, CD 3.1 m/3.4 m, 237 y, Fl. Jan. to Apr. Cracks on bark, and affected by nearby tree, its canopy leans to one side. Grows well. Managed and maintained by Chongqing Shaping Park Administrative Office.

古 11：红佛鼎
C. japonica 'Hongfoding'

位于重庆市沙坪公园办公楼前，N29°33′13″，E106°27′06″，海拔 244.5 m。株高 2.9 m，树干五分枝，基围 55.0 cm，冠幅 4.6/3.7 m，树龄 221 年，花期 1–4 月，长势良好。管护单位重庆市沙坪公园管理处。

Located in the front of Chongqing Shaping Park office, with latitude 29°33′13″ N, longitude 106°27′06″ E, elevation 244.5 m, tree H 2.9 m, five branches, CG 55.0 cm, CD 4.6 m/3.7 m, 221 y, Fl. Jan. to Apr. Grows well. Managed and maintained by Chongqing Shaping Park Administrative Office.

古 12: 白洋片
C. japonica 'Baiyangpian'

位于重庆市沙坪公园办公楼前，N29°33′13″，E106°27′06″，海拔244.5 m。株高3.1 m，基围44.0 cm，冠幅3.0 m，树龄187年，花期1–4月，长势良好。管护单位重庆市沙坪公园管理处。

Located in the front of Chongqing Shaping Park office, with latitude 29°33′13″ N, longitude 106°27′06″ E, elevation 244.5 m, tree H 3.1 m, CG 44.0 cm, CD 3.0 m, 187 y, Fl. Jan. to Apr. Grows well. Managed and maintained by Chongqing Shaping Park Administrative Office.

古 13：金顶大红
C. japonica 'Jinding Dahong'

位于重庆市沙坪公园办公楼前，N29°33′13″，E106°27′06″，海拔 244.7 m。株高 3.8 m，两分枝，基围 51.0 cm，冠幅 3.0 m，树龄 209 年，花期 1–4 月，长势良好。管护单位重庆市沙坪公园管理处。

Located in the front of Chongqing Shaping Park office, with latitude 29°33′13″ N, longitude 106°27′06″ E, elevation 244.7 m, tree H 3.8 m, two branches, CG 51.0 cm, CD 3.0 m, 209 y, Fl. Jan. to Apr. Grows well. Managed and maintained by Chongqing Shaping Park Administrative Office.

古14：白宝塔
C. japonica 'Baibaota'

位于重庆市沙坪公园办公楼前，N29°33′13″，E106°27′06″，海拔244.6 m。株高3.8 m，基围39.0 cm，冠幅4.0/3.5 m，树龄172年，花期2–4月，长势良好。管护单位重庆市沙坪公园管理处。

Located in the front of Chongqing Shaping Park office, with latitude 29°33′13″ N, longitude 106°27′06″ E, elevation 244.6 m, tree H 3.8 m, CG 39.0 cm, CD 4.0 m/3.5 m, 172 y, Fl. Feb. to Apr. Grows well. Managed and maintained by Chongqing Shaping Park Administrative Office.

重庆
抗战遗址博物馆

INTRODUCTION TO
CHONGQING HISTORIC SITES MUSEUM OF THE WAR OF
RESISTANCE AGAINST JAPAN

重庆抗战遗址博物馆位于重庆市南岸区南山生态带黄山景区内，占地面积 18.7hm²。20 世纪 20 年代，重庆白礼洋行买办黄云阶在此购建别墅取名黄家花园，又名黄山。抗战时期，国民政府迁都重庆，作为蒋介石黄山官邸。景区内有树龄百年以上古树和近百年大树众多，现有川山茶大树古树 13 棵。由于景区改造，埋土过深，排水不畅等原因，多数古茶树长势较弱，急需采取复壮措施，加强保护。

Chongqing Historic Sites Museum of the War of Resistance Against Japan is in the Huangshan Mountain Area, Nanshan Ecological Belt, Nan'an District, Chongqing City, covering an area of 18.7 hm². In the 1920s, Huang Yunjie, the comprador of Baili International Company in Chongqing, bought and built a villa here and named it Huangjia Garden, also known as Huangshan Mountain. During the Anti-Japanese War, the National Government of the Republic of China moved to Chongqing, and the Huangjia Garden served as the official residence of Chiang Kai-shek. In Huangshan Mountain area, there are plenty of trees over 100 years old and nearly 100 years old, including 13 Sichuan ancient and big camellia trees. Due to the reconstruction of the scenic area, the buried soil around those camellia trees are too deep that lead to poor drainage, which hinders the growth of nearby camellia trees. It is now urgent to protect and re-vitalize those trees.

古1：七心红玛瑙
C. japonica 'Qixinhong Manao'

位于重庆抗战遗址博物馆邮局停车场处，N29°34′13″，E106°36′54″，海拔508.5 m。株高2.6 m，基围52.0 cm，冠幅1.8 m，树龄109年，花期1–4月。植株中下部枝丫修剪痕迹明显，截枝处空腐严重，多枝已枯死，长势差。管护单位重庆抗战遗址博物馆。

Located in the parking lot of Chongqing Historic Sites Museum of the War of Resistance against Japan, with latitude 29°34′13″ N, longitude 106°36′54″ E, elevation 508.5 m, tree H 2.6 m, CG 52.0 cm, CD 1.8 m, 109 y, Fl. Jan. to Apr. The middle and lower branches have obvious pruning marks. Severe hollow rot at the cut branch with multiple dead twigs. Grows weakly. Managed and maintained by Chongqing Historic Sites Museum of the War of Resistance against Japan.

古2：紫金冠
C. japonica 'Zijinguan'

位于重庆抗战遗址博物馆邮局停车场处，N29°34′13″，E106°36′54″，海拔504.2 m。株高3.5 m，两分枝，冠幅2.3/2.0 m，树龄104年，花期1–4月。整株主干已全部空腐，仅剩一张树皮支撑，长势差。管护单位重庆抗战遗址博物馆。

Located in the parking lot of Chongqing Historic Sites Museum of the War of Resistance against Japan, with latitude 29°34′13″ N, longitude 106°36′54″ E, elevation 504.2 m, tree H 3.5 m, two branches, CD 2.3 m/2.0 m, 104 y, Fl. Jan. to Apr. The main trunk has been completely rotted, only with bark left for survival. Grows weakly. Managed and maintained by Chongqing Historic Sites Museum of the War of Resistance against Japan.

古 3: 七心红
C. japonica 'Qixinhong'

位于重庆抗战遗址博物馆办公室右侧，N 29°34′12″, E106°36′54″, 海拔505.1 m。株高4.3 m, 三分枝, 最大两分枝44.0/36.0 cm, 冠幅6.0 m, 树龄109年, 花期1–4月, 长势良好。管护单位重庆抗战遗址博物馆。

Located at the right of Chongqing Historic Sites Museum of the War of Resistance against Japan office, with latitude 29°34′12″ N, longitude 106°36′54″ E, elevation 505.1 m, tree H 4.3 m, three branches (the biggest two 44.0 cm/36.0 cm respectively), CD 6.0 m, 109 y, Fl. Jan. to Apr. Grows well. Managed and maintained by Chongqing Historic Sites Museum of the War of Resistance against Japan.

古 4：胭脂鳞
C. japonica 'Yanzhilin'

位于重庆市抗战遗址博物馆办公室右侧，N29°34′11″，E106°36′54″，海拔 505.4 m。株高 2.3 m，基围 50.0 cm，冠幅 3.2 m，树龄 104 年，花期 1–4 月。植株下部表皮受损，长势良好。管护单位重庆抗战遗址博物馆。

Located at the right of Chongqing Historic Sites Museum of the War of Resistance against Japan office, with latitude 29°34′11″ N, longitude 106°36′54″ E, elevation 505.4 m, tree H 2.3 m, CG 50.0 cm, CD 3.2 m, 104 y, Fl. Jan. to Apr. The base bark of the trunk was damaged. Grows well. Managed and maintained by Chongqing Historic Sites Museum of the War of Resistance against Japan.

古5：金盘荔枝
C. japonica 'Jinpan Lizhi'

位于重庆抗战遗址博物馆办公室右侧，N29°34′11″，E 106°36′53″，海拔 505.0 m。株高 5.3 m，基围 50.0 cm，冠幅 3.2/3.8 m，树龄 104 年，花期 1–4 月，长势一般。管护单位重庆抗战遗址博物馆。

Located at the right of Chongqing Historic Sites Museum of the War of Resistance against Japan office, with latitude 29°34′11″ N, longitude 106°36′53″ E, elevation 505.0 m, tree H 5.3 m, CG 50.0 cm, CD 3.2 m/3.8 m, 104 y, Fl. Jan. to Apr. Grows in general. Managed and maintained by Chongqing Historic Sites Museum of the War of Resistance against Japan.

古 6：七心红
C. japonica 'Qixinhong'

位于重庆抗战遗址博物馆第三会议室外，N29°34′11″，E106°36′55″，海拔 503.0 m。株高 3.0 m，基围 60.0 cm，冠幅 4.3/3.9 m，树龄 130 年，花期 1–4 月。树干基部有空腐和瘤状凸起，长势差。管护单位重庆抗战遗址博物馆。

Located outside the third conference room of Chongqing Historic Sites Museum of the War of Resistance against Japan, with latitude 29°34′11″ N, longitude 106°36′55″ E, elevation 503.0 m, tree H 3.0 m, CG 60.0 cm, CD 4.3 m/3.9 m, 130 y, Fl. Jan. to Apr. Hollow rot and tumor-like protrusion can be observed at the base of the trunk. Grows weakly. Managed and maintained by Chongqing Historic Sites Museum of the War of Resistance against Japan.

古7：花洋红
C. japonica 'Hua Yanghong'

位于重庆抗战遗址博物馆第三会议室外，N29°34′11″，E106°36′55″，海拔 503.5 m。株高 2.5 m，基围 62.0 cm，冠幅 2.1/1.6 m，树龄 135 年，花期 1–4 月。基部两分枝，主枝已枯死，留下空腐痕迹，长势差。管护单位重庆抗战遗址博物馆。

Located outside the third conference room of Chongqing Historic Sites Museum of the War of Resistance against Japan, with latitude 29°34′11″ N, longitude 106°36′55″ E, elevation 503.5 m, tree H 2.5 m, CG 62.0 cm, CD 2.1 m/1.6 m, 135 y, Fl. Jan. to Apr. Two branches grow from the trunk base, while the main trunk is dead. Grows weakly. Managed and maintained by Chongqing Historic Sites Museum of the War of Resistance against Japan.

古8：白洋片
C. japonica 'Baiyangpian'

位于重庆抗战遗址博物馆第三会议室外路口处，N29°34′10″，E106°36′55″，海拔505.2 m。株高3.0 m，冠幅3.1/2.6 m，树龄135年，花期2–4月。树干基部和其中一个分枝空腐，仅剩一张树皮支撑，长势良好。管护单位重庆抗战遗址博物馆。

Located at the intersection outside the third conference room of Chongqing Historic Sites Museum of the War of Resistance against Japan, with latitude 29°34′10″ N, longitude 106°36′55″ E, elevation 505.2 m, tree H 3.0 m, CD 3.1 m/2.6 m, 135 y, Fl. Feb. to Apr. The trunk base and one of the branches are hollowly rotted, only the bark left. Grows well. Managed and maintained by Chongqing Historic Sites Museum of the War of Resistance against Japan.

鹅岭公园

INTRODUCTION TO
ELING PARK

鹅岭公园位于重庆市渝中区鹅岭正街，于清宣统元年（1909年），由云南昭通府恩安县盐商李耀庭、四川通省劝业道重庆府劝业员李龢阳父子修建，时称"宜园"，后改称"礼园"。抗战时期，蒋介石和宋美龄曾在园中"飞阁"居住。园内地形较为复杂，植物种类丰富，现有川山茶古树3棵。2022年入选重庆市首批历史名园。

Eling Park is located on Eling Street, Yuzhong District, Chongqing. It was built in the first year of Xuantong period of the Qing Dynasty (1909) by Li Yaoting, a salt merchant from En'an County, Zhaotong Prefecture, Yunnan Province, and Li Heyang and his son. It was called "Yi Yuan" originally and later was renamed to "Li Yuan". During the Anti-Japanese War, Chiang Kai-shek and Soong May-ling lived in "Feige Pavilion" located inside the garden for a period. The garden features complicated topography and contains a wild profusion of plants and trees. There are 3 Sichuan ancient camellia trees. In 2022, the park was selected into the first batch of famous historic parks in Chongqing.

古1：七心红
C. japonica 'Qixinhong'

位于重庆市鹅岭公园盆景园门前，N29°33′06″，E106°31′51″，海拔 342.8 m。株高 3.1 m，基围 52.0 cm，冠幅 3.6 m，树龄 162 年，花期 1–4 月，两分枝，长势良好。管护单位重庆市鹅岭公园管理处。

Located in front of the Bonsai Garden of Chongqing Eling Park, with latitude 29°33′06″ N, longitude 106°31′51″ E, elevation 342.8 m, tree H 3.1 m, CG 52.0 cm, CD 3.6 m, 162 y, Fl. Jan. to Apr. Two branches. Grows well. Managed and maintained by Chongqing Eling Park Administrative Office.

古 2：七心红
C. japonica 'Qixinhong'

位于重庆市鹅岭公园两江亭下方，广星园门前，N29°33′09″，E106°32′04″，海拔 358.1 m。株高 3.8 m，基围 40.0 cm，冠幅 3.3 m，树龄 130 年，花期 1–4 月，长势良好。管护单位重庆市鹅岭公园管理处。

Located in front of the Guangxing Yard under the Liangjiang Pavilion, Chongqing Eling Park, with latitude 29°33′09″ N, longitude 106°32′04″ E, elevation 358.1 m, tree H 3.8 m, CG 40.0 cm, CD 3.3 m, 130 y, Fl. Jan. to Apr. Grows well. Managed and maintained by Chongqing Eling Park Administrative Office.

古3：紫金冠
C. japonica 'Zijinguan'

位于重庆市鹅岭公园两江亭下方，广星园门前，N29°33'08″，E106°32'04″，海拔326.9 m。株高5.0 m，基围60.0 cm，冠幅3.6 m，树龄183年，花期2–4月，两分枝，长势良好。管护单位重庆市鹅岭公园管理处。

Located in front of the Guangxing Yard under the Liangjiang Pavilion, Chongqing Eling Park, with latitude 29°33'08″ N, longitude 106°32'04″ E, elevation 326.9 m, tree H 5.0 m, CG 60.0 cm, CD 3.6 m, 183 y, Fl. Feb. to Apr. Two branches. Grows well. Managed and maintained by Chongqing Eling Park Administrative Office.

北碚公园

INTRODUCTION TO
BEIBEI PARK

北碚公园历史悠久，是著名民族实业家卢作孚先生于 1930 年倡议兴建的，地处北碚区老城区中心制高点的火焰山，现有面积 7.43hm^2。园内大树参天，花木繁茂，是富有历史文化和山地园林景观特色的城市公园，公园内现有川山茶大树古树 3 株。2022 年入选重庆市首批历史名园。

Beibei Park has a long history. It was proposed by Mr. Lu Zuofu, a famous national industrialist, and was constructed in 1930. The park is located on the Flaming Mountain, which is the Commanding elevation of the old town center of Beibei District. The existing area is 7.43 hm^2. There are plenty of towering trees and flowers. It is an urban park with rich history, culture and lush distinctive landscape of mountains and gardens. There are 3 Sichuan ancient and big camellia trees in the park. In 2022, the park was selected into the first batch of famous historic parks in Chongqing.

古 1：紫金冠
C. japonica 'Zijinguan'

位于重庆市北碚公园茶园下方，作孚园字碑前，N29°49′51″，E106°26′16″，海拔 177.7 m。株高 4.6 m，基围 58.0 cm，冠幅 4.2/2.8 m，树龄 127 年，花期 1–4 月。受一旁的桂花树影响，树冠偏向一边，现长势良好。管护单位重庆市北碚公园管理处。

Located in front of the Zuofu Monument, under the Tea Garden, Chongqing Beibei Park, with latitude 29°49′51″ N, longitude 106°26′16″ E, elevation 177.7 m, tree H 4.6 m, CG 58.0 cm, CD 4.2 m/2.8 m, 127 y, Fl. Jan. to Apr. Affected by nearby Osmanthus tree, its canopy leans to one side. Grows well. Managed and maintained by Chongqing Beibei Park Administrative Office.

花卉园

INTRODUCTION TO HUAHUI PARK

古1: 花洋红
C. japonica 'Hua Yanghong'

位于重庆市花卉园办公室前,N29°35′03″,E106°30′32″,海拔251.2 m。株高3.4 m,基围39.0 cm,冠幅2.7 m,树龄128年,花期1-4月。树干已空腐,受周围大树影响,太荫蔽,长势差,需加强养护。管护单位重庆市花卉园管理处。

Located in front of the Chongqing Huahui Park office, with latitude 29°35′03″ N, longitude106°30′32″ E, elevation 251.2 m, tree H 3.4 m, CG 39.0 cm, CD 2.7 m, 128 y, Fl. Jan. to Apr. The trunk is hollowly rotted. Affected by the adjacent giant trees, it can't get enough sunlight and grows weakly. It needs better protection and management. Managed and maintained by Chongqing Huahui Park Administrative Office.

古 2：白洋片
C. japonica 'Baiyangpian'

位于重庆市花卉园办公室前，N29°35′03″，E106°30′33″，海拔 254.4 m。株高 5.2 m，基围 35.0 cm，冠幅 4.3/3.3 m，树龄 117 年，花期 2–4 月，长势一般。管护单位重庆市花卉园管理处。

Located in front of the Chongqing Huahui Park office, with latitude 29°35′03″ N, longitude 106°30′33″ E, elevation 254.4 m, tree H 5.2 m, CG 35.0 cm, CD 4.3 m/3.3 m, 117 y, Fl. Feb. to Apr. Grows in general. Managed and maintained by Chongqing Huahui Park Administrative Office.

巴南区
华林路

*INTRODUCTION TO
HUALIN ROAD, BANAN DISTRICT*

古 1：红绣球
C. japonica 'Hongxiuqiu'

位于重庆市巴南区华林路918号，N29°26′36″，E106°38′22″，海拔306.7 m。株高6.4 m，基围73.0 cm，冠幅5.1 m，树龄144年，花期1–4月，长势良好。私家管护。

Located at No. 918 Hualin road, Banan District, Chongqing, with latitude 29°26′36″ N, longitude 106°38′22″ E, elevation 306.7 m, tree H 6.4 m, CG 73.0 cm, CD 5.1 m, 144 y, Fl. Jan. to Apr. Grows well. Private management.

古 2：红绣球
C. japonica 'Hongxiuqiu'

位于重庆市巴南区华林路 918 号，N29°25′42″，E106°40′11″，海拔 614.5 m。株高 4.6 m，基围 70.7 cm，冠幅 4.0 m，树龄 144 年，花期 1–4 月，长势良好。私家管护。

Located at No. 918 Hualin road, Banan District, Chongqing, with latitude 29°25′42″ N, longitude 106°40′11″ E, elevation 614.5 m, tree H 4.6 m, CG 70.7 cm, CD 4.0 m, 144 y, Fl. Jan. to Apr. Grows well. Private management.

缙云寺

INTRODUCTION TO
JINYUN TEMPLE

缙云寺为全国唯一的迦叶道场，位于重庆缙云山国家级自然保护区内。寺庙始建于南朝刘宋景平元年（423年），经历史变迁，明末清初寺毁于火灾，清康熙二十二年（1683年）由破空和尚主持修复。寺庙周围现有川山茶大树古树3株。

Jinyun Temple, the only Kasyapa Dojo in China, is in the National Nature Reserve of Jinyun Mountain in Chongqing. The temple was built in the first year (423) of Jingping period in the Southern Dynasty, experienced historical changes of dynasties. The temple was destroyed by fire in the late Ming and early Qing dynasties. In the 22nd year of Kangxi period of the Qing Dynasty (1683), it was repaired by the monk Pokong. There are three Sichuan ancient and big camellia trees around it.

古1：白洋片
C. japonica 'Baiyangpian'

位于重庆市北碚区缙云山国家自然保护区缙云寺庙外篮球场上方路口处，N29°50′17″，E106°23′24″，海拔727.6 m。株高3.2 m，基围54.0 cm，冠幅3.6 m，树龄117年，花期2-4月，长势差。管护单位重庆缙云山自然保护区。

Located at the upper crossroad of the basketball court of Jinyun mountain National Nature Reserve, Beibei District, Chongqing, with latitude 29°50′17″ N, longitude 106°23′24″ E, elevation 727.6 m, tree H 3.2 m, CG 54.0 cm, CD 3.6 m, 117 y. Fl. Feb. to Apr. Grows weakly. Managed and maintained by Chongqing Jinyun Mountain Nature Reserve.

古 2：七心红
C. japonica 'Qixinhong'

位于重庆市北碚区缙云山国家自然保护区缙云寺庙外篮球场上方路口处，N29°50′17″，E106°23′24″，海拔727.2 m。株高2.9 m，基围60.0 cm，冠幅2.8 m，树龄132年，花期1–4月，长势差。管护单位重庆缙云山自然保护区。

Located at the upper crossroad of the basketball court of Jinyun mountain National Nature Reserve, Beibei District, Chongqing, with latitude 29°50′17″ N, longitude 106°23′24″ E, elevation 727.2 m, tree H 2.9 m, CG 60.0 cm, CD 2.8 m, 132 y. Fl. Jan. to Apr. Grows weakly. Managed and maintained by Chongqing Jinyun Mountain Nature Reserve.

重庆市聚奎中学校

INTRODUCTION TO
CHONGQING JUKUI MIDDLE SCHOOL

重庆市聚奎中学校位于重庆市江津区白沙镇黑石山村，是重庆市重点中学。该校从清代同治九年创办聚奎义塾开始，发展成聚奎书院，几经传承演变，延续至今已有一百五十年的历史。校园内保存有川山茶大树古树7棵。由于四合院天井空间狭小，植株缺乏光照，管理粗放，现普遍长势较弱、急需加强扶壮与保护措施。建议疏散周围植物，增加采光和土壤通透性，加强肥水管理和病虫害防治。

Chongqing Jukui Middle School is located in Heishishan Village, Baisha Town, Jiangjin District, Chongqing. It is one of the key middle school in Chongqing. The school is originally the Jukui School established in the ninth year of Tongzhi period in the Qing Dynasty, and later developed into Jukui Academy. After several inheritance and evolution, it continues to this day as Chongqing Jukui Middle School, with a history of 150 years. It preserves seven Sichuan ancient and big camellia trees. However, the growth of those camellia trees was impeded due to narrow space, lack of daylight and careless management. It is urgently to protect those trees for better growth and revitalization. It is suggested to reduce density of surrounding plants, increase daylight and soil permeability, enhance fertilizer and water management and consider pest control.

古1：金顶大红
C. japonica 'Jinding Dahong'

位于重庆市江津区聚奎中学党建办门前天井处，N29°02′32″，E106°08′12″，海拔271.6 m。株高4.6 m，基围75.0 cm，冠幅4.4/3.2 m，树龄169年，花期1—4月。长势一般，枝叶中等偏少，主杆基部有空腐，一枝枯死，树形严重偏冠。管护单位重庆市聚奎中学校。

Located at a patio in front of the Party Construction Office of Chongqing Jukui Middle School, Jiangjin District, Chongqing, with latitude 29°02′32″ N, longitude 106°08′12″ E, elevation 271.6 m, tree H 4.6 m, CG 75.0 cm, CD 4.4 m/3.2 m, 169 y, Fl. Jan. to Apr. It grows generally with medium-to-low amount of foliage. The trunk base has been hollowly rotted and one branch is dead. Managed and maintained by Chongqing Jukui Middle School.

古2：胭脂鳞
C. japonica 'Yanzhilin'

位于重庆市江津区聚奎中学党建办门前天井处，N29°02′31″，E106°08′12″，海拔268.0 m。株高3.9 m，基围55.0 cm，冠幅3.2/2.8 m，树龄117年，花期1-4月，长势良好。管护单位重庆市聚奎中学校。

Located at a patio in front of the Party Construction Office of Chongqing Jukui Middle School, Jiangjin District, Chongqing, with latitude 29°02′31″ N, longitude 106°08′12″ E, elevation 268.0 m, tree H 3.9 m, CG 55.0 cm, CD 3.2 m/2.8 m, 117 y, Fl. Jan. to Apr. Grows well. Managed and maintained by Chongqing Jukui Middle School.

古 3：川玛瑙
C. japonica 'Chuanmanao'

位于重庆市江津区聚奎中学党建办门前天井处，N29°02′31″，E106°08′12″，海拔 269.8 m。株高 5.2 m，两分枝，基围 52.0 cm，冠幅 2.1/2.6 m，树龄 109 年，花期 1–4 月。主干基部至其中一分枝空腐，只剩一张树皮，枝叶中等偏少，长势一般。管护单位重庆市聚奎中学校。

Located at a patio in front of the Party Construction Office of Chongqing Jukui Middle School, Jiangjin District, Chongqing, with latitude 29°02′31″ N, longitude 106°08′12″ E, elevation 269.8 m, tree H 5.2 m, two branches, CG 52.0 cm, CD 2.1 m/2.6 m, 109 y, Fl. Jan. to Apr. The trunk base and one branch are hollowly rotted, only bark left. It grows generally with medium-to-low amount of foliage. Managed and maintained by Chongqing Jukui Middle School.

重庆金佛山
国家级自然保护区

INTRODUCTION TO
CHONGQING GOLD BUDDHA MOUNTAIN
NATIONAL NATURE RESERVE

重庆金佛山国家级自然保护区位于重庆市南川区境内，有着"天然植物陈列馆"的美誉，在山中 167 km² 原始常绿林中，萃集着 237 科 2997 种植物。2017 年 4 月，植物专家在重庆南川金佛山南麓发现了疏齿大厂茶的新种群。40 余株野生古树茶集中生长在一公里范围中。在南川区三全镇黄草坪周边有上千株西南山茶古树。

Chongqing Gold Buddha Mountain National Nature Reserve, located in Nanchuan District, Chongqing, is also known as "Natural Plant Exhibition Hall". The Reserve contains a pristine evergreen forest over 167 km² in area, which preserves 237 families of various plants which include 2997 different species. In April 2017, botanists discovered a new type of wild *C. tachangensis* var. remotiserrata Chang at the southern foot. Around 40 wild *C.tachangensis* var. remotiserrata Chang grow adjacently within one kilometer. Meanwhile, thousands of *C. pitardii* are growing around the Huangcaoping in Sanquan Town, Nanchuan District.

古 1：西南白山茶
C. pitardii Coh. St. var. *alba* Chang

位于重庆市金佛山三泉镇三泉村大龙林，N29°03′16″，E107°12′09″，海拔1286.3 m。株高5.0 m，基围66.0 cm，两分枝，主干被砍后长出丛生新枝。冠幅2.0 m，树龄136年，花期11月至翌年3月。管护单位重庆市南川区三全镇林场。

Located in Dalonglin at Sanquan Village, Sanquan Town, Golden Buddha Mountain, Chongqing City, with latitude 29°03′16″ N, longitude 107°12′09″ E, elevation 1286.3 m, tree H 5.0 m, CG 66.0 cm, two branches. Several branches grow out after the cutting of main trunk. CD 2.0 m, 136 y, Fl. Nov. to Mar. Managed and maintained by Chongqing Nanchuan District Sanquan Forest Farm.

古 2：西南白山茶
C. pitardii Coh. St. var. *alba* Chang

位于重庆市金佛山三泉镇三泉村大龙林，N29°03′15″，E107°12′10″，海拔 1281.6 m。株高 5.0 m，基围 80.0 cm，冠幅 3.0 m，主干被砍后长出丛生新枝。树龄 152 年，花期 11 月至翌年 3 月。管护单位重庆市南川区三全镇林场。

Located in Dalonglin at Sanquan Village, Sanquan Town, Golden Buddha Mountain, Chongqing City, with latitude 29°03′15″ N, longitude 107°12′10″ E, elevation 1281.6 m, tree H 5.0 m, CG 80.0 cm, CD 3.0 m, 152 y, Fl. Nov. to Mar. Several branches grow out after the cutting of main trunk. Managed and maintained by Chongqing Nanchuan District Sanquan Forest Farm.

古3：西南白山茶
C. pitardii Coh. St. var. *alba* Chang

位于重庆市金佛山三泉镇三泉村大龙林，N29°03′15″，E107°12′10″，海拔 1283.2 m。株高 6.6 m，基围 96.0 cm，原两分枝，被砍后长出六分枝，冠幅 3.5 m，树龄 171 年，花期 11 月至翌年 3 月。管护单位重庆市南川区三全镇林场。

Located in Dalonglin at Sanquan Village, Sanquan Town, Golden Buddha Mountain, Chongqing City, with latitude 29°03′15″ N, longitude 107°12′10″ E, elevation 1283.2 m, tree H 6.6 m, CG 96.0 cm, six branches grow after the cutting of the original two branches. CD 3.5 m, 171 y, Fl. Nov. to Mar. Managed and maintained by Chongqing Nanchuan District Sanquan Forest Farm.

古 4：西南白山茶
C. pitardii Coh. St. var. *alba* Chang

位于重庆市金佛山三泉镇三泉村大龙林，N29°03′14″，E107°12′09″，海拔 1278.7 m。株高 6.0 m，基围 90.0 cm，主干被砍后长出四分枝，冠幅 3.8 m，树龄 164 年，花期 11 月至翌年 3 月。管护单位重庆市南川区三全镇林场。

Located in Dalonglin at Sanquan Village, Sanquan Town, Golden Buddha Mountain, Chongqing City, with latitude 29°03′14″ N, longitude 107°12′09″ E, elevation 1278.7 m, tree H 6.0 m, CG 90.0 cm, CD 3.8 m, 164 y, Fl. Nov. to Mar. Four branches grow after the cutting of main trunk. Managed and maintained by Chongqing Nanchuan District Sanquan Forest Farm.

古5：西南红山茶
C. pitardii Coh. St. var. Pitardii Sealy

位于重庆市金佛山三泉镇三泉村大龙林，N29°02′49″，E109°10′54″，海拔1728.6 m。株高8.0 m，基围40.0 cm，冠幅4.0 m，树龄105年，花期11月至翌年3月。管护单位重庆市南川区三全镇林场。

Located in Dalonglin at Sanquan Village, Sanquan Town, Golden Buddha Mountain, Chongqing City, with latitude 29°02′49″ N, longitude 109°10′54″ E, elevation 1728.6 m, tree H 8.0 m, CG 40.0 cm, CD 4.0 m, 105 y, Fl. Nov. to Mar. Managed and maintained by Chongqing Nanchuan District Sanquan Forest Farm.

古 6：西南红山茶
C. pitardii Coh. St. var. *Pitardii* Sealy

位于重庆市金佛山三泉镇三泉村大龙林，N29°02′50″，E109°10′55″，海拔 1732.5 m。株高 14.0 m，基围 100.0 cm，冠幅 6.5 m，树龄 176 年，花期 11 月至翌年 3 月。管护单位重庆市南川区三全镇林场。

Located in Dalonglin at Sanquan Village, Sanquan Town, Golden Buddha Mountain, Chongqing City, with latitude 29°02′50″ N, longitude 109°10′55″ E, elevation 1732.5 m, tree H 14.0 m, CG 100.0 cm, CD 6.5 m, 176 y, Fl. Nov. to Mar. Managed and maintained by Chongqing Nanchuan District Sanquan Forest Farm.

古 7：西南红山茶
C. pitardii Coh. St. var. *Pitardii* Sealy

位于重庆市金佛山三泉镇三泉村大龙林，N29°02′50″，E109°10′55″，海拔 1735.1 m。株高 9.0 m，基围 95.0 cm，两分枝，最大分枝 68.0 cm，冠幅 5.0/4.3 m，树龄 170 年，花期 11 月至翌年 3 月。管护单位重庆市南川区三全镇林场。

Located in Dalonglin at Sanquan Village, Sanquan Town, Golden Buddha Mountain, Chongqing City, with latitude 29°02′50″ N, longitude 109°10′55″ E, elevation 1735.1 m, tree H 9.0 m, CG 95.0 cm, two branches (the biggest one 68.0 cm), CD 5.0 m/4.3 m, 170 y, Fl. Nov. to Mar. Managed and maintained by Chongqing Nanchuan District Sanquan Forest Farm.

重庆武陵山
国家森林公园

INTRODUCTION TO CHONGQING
WULING MOUNTAIN NATIONAL FOREST PARK

重庆武陵山国家森林公园位于重庆市涪陵区城东南约 55 km 的大木乡境内，拥有原始森林 667 km²，被誉为重庆最大的肺叶。景区内峡谷、地缝、森林、花卉、民俗等旅游资源丰富，核心景区海拔高 2000 m，遍布原始森林，野生动物，奇珍异果，有"清凉仙境"，"童话王国"的美誉。在重庆市武隆区、涪陵区、丰都县交界处分布有几千株野生山茶古树。

Chongqing Wuling Mountain National Forest Park is located in Damu Town, which is around 55 km away southeast of Fuling District, Chongqing. It contains virgin forests over 667 km² in area, thus it is also known as the largest "lung lobe" in Chongqing. The scenic area is rich in tourism resources such as canyons, ground crevices, forests, flowers, and folk customs. The main scenic spot is at altitude of 2,000 m , and is covered with virgin forests, various wild animals and exotic fruits, thus is also known as "Cool Wonderland" and "Fairy Tales". There are thousands of ancient wild camellia trees at the junction of Wulong District, Fuling District and Fengdu County in Chongqing.

古 1: 西南红山茶
C. pitardii Coh. St. var. *Pitardii* Sealy

位于重庆市涪陵区双河乡，N29°33'50"，E107°39'10"，海拔1797.0 m。株高4.8 m，基围134.0 cm，五分枝，最大一枝基围54.0 cm，冠幅4.6 m，树龄218年，花期11月至翌年4月，长势良好。管护单位重庆市武隆双河乡擂子湾林场。

Located at Shuanghe Village, Fuling District, Chongqing, with latitude 29°33'50" N, longitude 107°39'10" E, elevation 1797.0 m, tree H 4.8 m, CG 134.0 cm, five branches (the biggest one 54.0 cm), CD 4.6 m, 218 y, Fl. Nov. to Apr. Grows well. Managed and maintained by Leiziwan Forest Farm of Shuanghe Village.

古 2：西南红山茶
C. pitardii Coh. St. var. *Pitardii* Sealy

位于重庆市涪陵区双河乡，N29°35'16"，E107°40'06"，海拔 1784.2 m。株高 7.6 m，基围 85.0 cm，四分枝，冠幅 6.0 m，树龄 158 年，花期 11 月至翌年 4 月，长势良好。管护单位重庆市武隆双河乡擂子湾林场。

Located at Shuanghe Village, Fuling District, Chongqing, with latitude 29°35'16" N, longitude 107° 40'06" E, elevation 1784.2 m, tree H 7.6 m, CG 85.0 cm, four branches, CD 6.0 m, 158 y, Fl. Nov. to Apr. Grows well. Managed and maintained by Leiziwan Forest Farm of Shuanghe Village.

古3：西南红山茶
C. pitardii Coh. St. var. *Pitardii* Sealy

位于重庆市涪陵区双河乡，N29°35'16"，E107°40'07"，海拔1790.0 m。株高7.5 m，基围65.0 cm，冠幅4.5 m，树龄134年，花期11月至翌年4月，主干已死，剩下侧枝，长势良好。管护单位重庆市武隆双河乡擂子湾林场。

Located at Shuanghe Village, Fuling District, Chongqing, with latitude 29°35'16" N, longitude 107°40'07" E, elevation 1790.0 m, tree H 7.5 m, CG 65.0 cm, CD 4.5 m, 134 y, Fl. Nov. to Apr. Grows well. Managed and maintained by Leiziwan Forest Farm of Shuanghe Village.

古 4：西南红山茶
C. pitardii Coh. St. var. *Pitardii* Sealy

位于重庆市涪陵区双河乡，N29°35'16"，E107°40'04"，海拔 1763.0 m。株高 4.5 m，基围 65.0 cm，六分枝，主干有被砍过的痕迹，分枝是后长出来的，冠幅 6.0 m，树龄 134 年，花期 11 月至翌年 4 月，长势良好。管护单位重庆市武隆双河乡擂子湾林场。

Located at Shuanghe Village, Fuling District, Chongqing, with latitude 29°35'16" N, longitude 107°40'04" E, elevation 1763.0 m, tree H 4.5 m, CG 65.0 cm, six branches, CD 6.0 m, 134 y, Fl. Nov. to Apr. Grows well. Managed and maintained by Leiziwan Forest Farm of Shuanghe Village.

古 5：西南红山茶
C. pitardii Coh. St. var. *Pitardii* Sealy

位于重庆市涪陵区双河乡，N29°34'47"，E107°38'47"，海拔 1693.0 m。株高 6.0 m，基围 137.0 cm，六分枝，最大一枝基围 40.0 cm，冠幅 3.5 m，树龄 219 年，花期 11 月至翌年 4 月，长势良好。管护单位重庆市武隆双河乡擂子湾林场。

Located at Shuanghe Village, Fuling District, Chongqing, with latitude 29°34'47" N, longitude 107°38'47" E, elevation 1693.0 m, tree H 6.0 m, CG 137.0 cm, six branches (the biggest one 40.0 cm), CD 3.5 m, 219 y, Fl. Nov. to Apr. Grows well. Managed and maintained by Leiziwan Forest Farm of Shuanghe Village.

古 6：西南红山茶
C. pitardii Coh. St. var. Pitardii Sealy

位于重庆市涪陵区双河乡，N29°34'50"，E107°38'45"，海拔 1661.5 m。株高 5.0 m，基围 145.0 cm，八分枝，最大一枝基围 54.0 cm，冠幅 6.0 m，树龄 229 年，花期 11 月至翌年 4 月，长势良好。管护单位重庆市武隆双河乡擂子湾林场。

Located at Shuanghe Village, Fuling District, Chongqing, with latitude 29°34'50" N, longitude 107°38'45" E, elevation 1661.5 m, tree H 5.0 m, CG 145.0 cm, eight branches (the biggest one 54.0 cm), CD 6.0 m, 229 y, Fl. Nov. to Apr. Grows well. Managed and maintained by Leiziwan Forest Farm of Shuanghe Village.

古 7：西南红山茶
C. pitardii Coh. St. var. *Pitardii* Sealy

位于重庆市涪陵区双河乡，N29°34'50"，E107°38'45"，海拔 1661.7 m。株高 3.5 m，基围 67.0 cm，四分枝，最大一枝基围 54.0 cm，冠幅 3.5 m，树龄 137 年，花期 11 月至翌年 4 月，受环境影响树冠偏冠严重，长势良好。管护单位重庆市武隆双河乡擂子湾林场。

Located at Shuanghe Village, Fuling District, Chongqing, with latitude 29°34'50" N, longitude 107°38'45" E, elevation 1661.7 m, tree H 3.5 m, CG 67.0 cm, four branches (the biggest one 54.0 cm), CD 3.5 m, 137 y, Fl. Nov. to Apr. Affected by environment, its canopy leans to one side. Grows well. Managed and maintained by Leiziwan Forest Farm of Shuanghe Village.

古8：西南红山茶
C. pitardii Coh. St. var. *Pitardii* Sealy

位于重庆市涪陵区双河乡，N29°34'50"，E107°38'47"，海拔1714.9 m。株高5.0 m，基围110.0 cm，五分枝，冠幅4.0 m，树龄187年，花期11月至翌年4月，长势良好。管护单位重庆市武隆双河乡擂子湾林场。

Located at Shuanghe Village, Fuling District, Chongqing, with latitude 29°34'50" N, longitude 107°38'47" E, elevation 1714.9 m, tree H 5.0 m, CG 110.0 cm, five branches, CD 4.0 m, 187 y, Fl. Nov. to Apr. Grows well. Managed and maintained by Leiziwan Forest Farm of Shuanghe Village.

古 9：西南红山茶
C. pitardii Coh. St. var. *Pitardii* Sealy

位于重庆市涪陵区双河乡，N29°34'50"，E107°38'45"，海拔 1661.5 m。株高 5.0 m，基围 112.0 cm，七分枝，冠幅 3.0 m，树龄 190 年，花期 11 月至翌年 4 月，长势良好。管护单位重庆市武隆双河乡擂子湾林场。

Located at Shuanghe Village, Fuling District, Chongqing, with latitude 29°34'50" N, longitude 107°38'45" E, elevation 1661.5 m, tree H 5.0 m, CG 112.0 cm, seven branches, CD 3.0 m, 190 y, Fl. Nov. to Apr. Grows well. Managed and maintained by Leiziwan Forest Farm of Shuanghe Village.

古 10：西南红山茶
C. pitardii Coh. St. var. *Pitardii* Sealy

位于重庆市涪陵区双河乡，N29°34'50"，E107°38'47"，海拔 1727.9 m。株高 5.0 m，基围 105.0 cm，两分枝，冠幅 5.0 m，树龄 182 年，花期 11 月至翌年 4 月，长势良好。管护单位重庆市武隆双河乡擂子湾林场。

Located at Shuanghe Village, Fuling District, Chongqing, with latitude 29°34'50" N, longitude 107°38'47" E, elevation 1727.9 m, tree H 5.0 m, CG 105.0 cm, two branches, CD 5.0 m, 182 y, Fl. Nov. to Apr. Grows well. Managed and maintained by Leiziwan Forest Farm of Shuanghe Village.

古 11：西南红山茶
C. pitardii Coh. St. var. *Pitardii* Sealy

位于重庆市涪陵区双河乡，N29°33'50"，E107°39'10"，海拔 1725.6 m。株高 4.0 m，基围 45.0 cm，冠幅 2.0 m，树龄 111 年，花期 11 月至翌年 4 月，长势良好。管护单位重庆市武隆双河乡擂子湾林场。

Located at Shuanghe Village, Fuling District, Chongqing, with latitude 29°33'50" N, longitude 107°39'10" E, elevation 1725.6 m, tree H 4.0 m, CG 45.0 cm, CD 2.0 m, 111 y, Fl. Nov. to Apr. Grows well. Managed and maintained by Leiziwan Forest Farm of Shuanghe Village.

古 12：西南红山茶
C. pitardii Coh. St. var. *Pitardii* Sealy

位于重庆市涪陵区双河乡，N29°34' 51"，E107°38'48"，海拔 1740.3 m。株高 4.5 m，基围 98.0 cm，五分枝，冠幅 5.5 m，树龄 173 年，花期 11 月至翌年 4 月，长势良好。管护单位重庆市武隆双河乡擂子湾林场。

Located at Shuanghe Village, Fuling District, Chongqing, with latitude 29°34'51" N, longitude 107°38'48" E, elevation 1740.3 m, tree H 4.5 m, CG 98.0 cm, five branches, CD 5.5 m, 173 y, Fl. Nov. to Apr. Grows well. Managed and maintained by Leiziwan Forest Farm of Shuanghe Village.

彭水苗族土家族自治县

**INTRODUCTION TO
PENGSHUI MIAO AND
TUJIA AUTONOMOUS COUNTY**

彭水苗族土家族自治县位于重庆市东南部，处武陵山区，居乌江下游，环境优越，资源丰富，是中国油茶基地重点县之一。在海拔 800~1500 m 山上分布有数千株西南山茶古树，主要种类有西南红山茶和多变性西南红山茶。其中西南红山茶主要分布在彭水东北方向的龙射镇、太原镇、棣棠乡等；多变性西南红山茶主要分布在彭水中部、西南方向的靛水街道、新田镇、岩东乡等地。目前发现集中成片规模最大的在新田镇，面积超过 6 hm^2。

Pengshui Miao and Tujia Autonomous County is located in the southeast of Chongqing, in the Wuling Mountains and the lower reaches of the Wujiang River. With superior environment and rich resources, it is one of the key counties of China's camellia base.

There are thousands of ancient camellia trees distributed on mountains at an altitude of 800 - 1500 meters, which mainly include *C. pitardii* Coh. St. var. *pitardii* Sealy and *C. pitardii* var. Compressa. Among them, *C. pitardii* Coh. St. var. *pitardii* Sealy mainly distribute in northeast of Pengshui county, such as Longshe Town, Taiyuan Town, Ditang Village, and etc., while *C. pitardii* var. Compressa mainly distributed in the middle and southwest of Pengshui autonomous county, including Dianshui Street, Xintian Town, Yandong Village and etc. Currently, it has been found that the greatest area of distribution of C. *pitardii* var. Compressa concentrated is more than 6 hm^2 in Xintian Town.

古1：西南红山茶
C. pitardii Coh. St. var. *Pitardii* Sealy

位于重庆市彭水县龙射镇胜利村一农户门前，N29°30′45″，E108°12′27″，海拔 1016.7 m。株高 4.5 m，基围 47.0 cm，冠幅 3.5 m，树龄 113 年，花期 11 月至翌年 3 月，现长势良好。为私人管护。

Located in front of a farmer's house in Shengli Village, Longshe Town, Pengshui County, Chongqing City, with latitude 29°30′45″ N, longitude 108°12′27″ E, elevation 1016.7 m, tree H 4.5 m, CG 47.0 cm, CD 3.5 m, 113 y, Fl. Nov. to Mar. Grows well. Private management.

古 2: 西南红山茶
C. pitardii Coh. St. var. *Pitardii* Sealy

位于重庆市彭水县龙射镇凉风村4组，N29°34′26″，E108°04′18″，海拔963.6 m。株高8.0 m，基围85.0 cm，冠幅6.5 m，树龄158年，花期11月至翌年3月。两分枝，被砍一枝，现长势良好。为私人管护。

Located in group 4 in Liangfeng Village, Longshe Town, Pengshui County, Chongqing City, with latitude 29°34′26″ N, longitude 108°04′18″ E, elevation 963.6 m, tree H 8.0 m, CG 85.0 cm, CD 6.5 m, 158 y, Fl. Nov. to Mar. Two branches and one was cut. Grows well currently. Private management.

古3：多变性西南红山茶
C. pitardii var. Compressa

位于重庆市彭水县岩东乡堰塘村一农地坟前，N29°17′31″，E108°15′47″，海拔990.1 m。株高6.0 m，基围78.0 cm，冠幅5.0 m，树龄150年，花期1-4月，因耕地已损伤一半根系，现长势良好。据彭水县林业局工作人员讲述，在海拔900~1500m有大量分布。为私人管护。

Located in the farmland in Yantang Village, Yandong Town, Pengshui County, Chongqing City, with latitude 29°17′31″ N, longitude 108°15′47″ E, elevation 990.1 m, tree H 6.0 m, CG 78.0 cm, CD 5.0 m, 150 y, Fl. Jan. to Apr. Half of root system has been damaged due to land cultivation. Grows well. According to the staff of Pengshui County Forestry Bureau, there are large distributions of this species at an altitude of 900-1500 m. Private management.

四面山
风景名胜区

INTRODUCTION TO
SIMIAN MOUNTAIN
NATIONAL NATURE RESERVE

位于重庆市江津区南部，地处云贵高原大娄山北坡余脉，是地球北纬 28° 线上仅存的面积最大、保护最完好的亚热带原始常绿阔叶林带，植被丰富，被誉为"天然物种基因库"。在海拔 1000~1800 m 分布有超过千株野生西南红山茶古树。

Chongqing Simian Mountain National Nature Reserve, located in the remnants of the northern slope of Dalou Mountain of Yunnan-Guizhou Plateau, is in the south of Jiangjin District, Chongqing. It is the largest and the best-preserved subtropical primitive evergreen broad-leaved forest belt on the 28° North latitude of the earth. The reserve contains a wild profusion and species of plants and trees, thus is also known as "Natural Species Gene Bank". It has been found that more than 1000 wild ancient trees of *C. pitardii* Coh.St. var. *Pitardii* Sealy distributed in the Reserve at an altitude of 1000-1800 meters.

古 1：西南红山茶
C. pitardii Coh. St. var. *Pitardii* Sealy

位于重庆市四面山镇头道河 1 社坪子头向姓村民家房前，N28°36′28″，E106°26′49″，海拔 1213.6 m。株高 3.3 m，基围 38.5 cm，冠幅 3.0/3.6 m，树龄 103 年，花期 11 月至翌年 3 月，长势良好。私人管护。

Located in front of Xiang's house, Simianshan Town, Chongqing City, with latitude 28°36′28″ N, longitude 106°26′49″ E, elevation 1213.6 m, tree H 3.3 m, CG 38.5 cm, CD 3.0 m/3.6 m, 103 y, Fl. Nov. to Mar. Grows well. Private management.

古2：西南红山茶
C. pitardii Coh. St. var. Pitardii Sealy

位于重庆市四面山镇头道河 1 社坪子头向姓村民家房前，N28°36′28″，E106°26′48″，海拔 1213.6 m。株高 2.5 m，基围 47.0 cm，三分枝，冠幅 2.6/3.2 m，树龄 113 年，花期 11 月至翌年 3 月，长势良好。私人管护。

Located in front of Xiang's house, Simianshan Town, Chongqing City, with latitude 28°36′28″ N, longitude 106°26′48″ E, elevation 1213.6 m, tree H 2.5 m, CG 47.0 cm, three branches, CD 2.6 m/3.2 m, 113 y, Fl. Nov. to Mar. Grows well. Private management.

古3：西南红山茶
C. pitardii Coh. St. var. Pitardii Sealy

位于重庆市四面山镇头道河1社坪子头向姓村民家房前，N28°36′28″，E106°26′48″，海拔1213.6 m。株高2.6 m，基围45.0 cm，冠幅2.0 m，树龄111年，花期11月至翌年3月。据主人介绍该树为三分枝，系其主干被砍后发出的侧枝，因其土埋太深，看不见主干，长势较差。私人管护。

Located in front of Xiang's house, Simianshan Town, Chongqing City, with latitude 28°36′28″ N, longitude 106°26′48″ E, elevation 1213.6 m, tree H 2.6 m, CG 45.0 cm, CD 2.0 m, 111 y, Fl. Nov. to Mar. According to its owner, it has three branches, which grown after the main trunk was cut. Filling soil around it is so deep that the main trunk cannot be observed. Grows weakly. Private management

以上三株茶花树是向姓村民十多年前从坪子头大山上挖回来栽种到房前，据他说山上有很多大树，独干和丛生的均有，丛生较多。因太大，不好搬运，未挖。

　　近年来因村民房屋改建，三株茶花均被填土太深，长势有不同程度的衰弱。

Those three camellia trees were dug up by villager Xiang from the mountain in Pingzitou over ten years ago and planted in front of his house. According to him, there are many big trees on the mountain, growing both in solitary and clumps, while relatively more growing clumps. He didn't dig those trees since they are too big to carry.

Due to the reconstruction of houses in the village, filling soil around those three camellia trees are too deep and limit their growth to some extent.

第二章
四川地区山茶古树资源分布

Chapter II
Distribution of Ancient Camellia Tree Resources in Sichuan region

　　四川地区川山茶古树包括栽培型古树和野生型古树，本书详细记载了四川地区 18 株栽培型山茶和 32 株野生型山茶。其中选取有代表性的单株 42 株古树进行详细介绍展示，栽培古树主要分布于私家园林和公园中，如都江堰离堆公园、雅安私人庭园，其他地点有零散分布。川山茶古树的品种主要为'红绣球''三学士''洋红''铁壳紫袍'等 7 个。部分川山茶栽培型古树因生长环境差，养护不力，其生长状况整体较差，多表现为叶片黄绿、枝叶稀疏，甚至伴有明显枯死枝。

　　野生型古树主要分布于雅安邛崃天台山、峨眉洪雅县等大山之中，海拔 800~1800 m 之间。以西南红山茶、西南白山茶和峨眉红山茶为主，生长健壮。

The ancient camellia trees in Sichuan include cultivated type and wild type. This book records 18 cultivated trees and 32 wild trees in Sichuan area. Among them, 42 typical camellia trees are selected with detailed introduction. The cultivated ancient trees are mainly distributed in private gardens and parks, such as Dujiangyan Lidui Park and Ya'an Private Garden, and are scattered in other locations. There are seven cultivars of Sichuan ancient camellia trees, including 'Hongxiuqiu' 'Sanxueshi' 'Yanghong' 'Tieke Zipao' and etc. Due to undesired environmental conditions as well as insufficient maintenance and caretaking, some cultivated trees grow weakly, mainly show as sparse branches and foliage, unhealthy yellow-green foliage and even with obvious dead twigs.

Wild-type camellia trees are mainly distributed in Qionglai Tiantai Mountain of Ya'an City and Hongya County of Emeishan City at altitude around 800-1800 m. Those wild-type camellia trees, mainly include *C. pitaradii* coh. St. var. *Pitardii* Sealy, *C. pitardii* var. *alba* Chang and *C. omeiensis* Chang, are growing vigorously.

都江堰离堆公园

INTRODUCTION TO
DUJIANGYAN LIDUI PARK

都江堰离堆公园，位于四川都江堰市城南公园路，是都江堰的入口部分，占地约 6 hm²。战国时郡守李冰修建都江堰时在玉垒山伸向岷江的石山上凿开了一个口子，称为宝瓶口，脱离部分的余脉石堆称为离堆。离堆园所在地古为"果园"，宋代名为"花洲"，清末改建为"桑园"，民国初年在园内设"蚕桑局"。民国十四年（1925年）改为公园，次年更名为离堆公园。园内古木桩头，奇花异卉，布局精巧，现有川山茶大树古树 7 株。

As the entrance of Dujiangyan, Dujiangyan Lidui Park located on Chengnan Gongyuan Road, covering an area of more than 6 hm². During the Warring States Period, when Li Bing, the county governor, built Dujiangyan. A gap was dug in the stone mountain extending from Yulei Mountain to the Minjiang River, which was later called Baopingkou, while the pile of removed stones was called Lidui. The location of Lidui Garden was called "Orchard" in ancient times, which was later called "Huazhou" in Song Dynasty, then rebuilt into "Mulberry Garden" in late Qing Dynasty. In the early years of the Republic of China, a "Sericulture Bureau" was set up in the Garden. In the 14th year of the Republic of China (1925), it was changed to a public park, and renamed as Lidui Park the following year. In the park, there are ancient bonsai, exotic flowers, and exquisite layout. There are 7 Sichuan ancient and big camellia trees preserved in the garden.

古 1：紫金冠
C. japonica 'Zijinguan'

盆景桩头。位于四川省都江堰离堆公园盆景园—清溪园内，N30°59′51″，E103°36′43″，海拔 670.4 m。株高 3.4 m，基围 60.0 cm，冠幅 2.1 m，树龄 164 年，花期 2–4 月。树干已枯腐，只剩下一张皮，长势差，建议地栽养护。管护单位都江堰旅游管理局离堆公园管理处。

Bonsai. Located in the Bonsai Garden of the Lidui Park, Dujiangyan City, Sichuan, with latitude 30°59′51″ N, longitude 103°36′43″ E, elevation 670.4 m, tree H 3.4 m, CG 60.0 cm, CD 2.1 m, 164 y, Fl. Feb. to Apr. The trunk is rotted, only bark left. Grows weakly. It is recommended to ground planting. Managed and maintained by Lidui Park Administrative Office of Dujiangyan Tourism Administration.

古 2: 三学士
C. japonica 'Sanxueshi'

位于四川省都江堰离堆公园盆景园—清溪园内，N30°59′51″，E103°36′43″，海拔 672.8 m。株高 4.3 m，基围 61.5 cm，冠幅 3.0/2.6 m，树龄 103 年，花期 3–5 月，长势良好。管护单位都江堰旅游管理局离堆公园管理处。

Located in the Bonsai Garden of the Lidui Park, Dujiangyan City, Sichuan, with latitude 30°59′51″ N, longitude 103°36′43″ E, elevation 672.8 m, tree H 4.3 m, CG 61.5 cm, CD 3.0 m/2.6 m, 103 y, Fl. Mar. to May. Grows well. Managed and maintained by Lidui Park Administrative Office of Dujiangyan Tourism Administration.

古 3: 铁壳紫袍
C. japonica 'Tieke Zipao'

位于四川省都江堰离堆公园盆景园—清溪园内，N30°59′52″，E103°36′42″，海拔 686.0 m。株高 4.1 m，基围 66.0 cm，冠幅 3.7/3.0 m，树龄 110 年，花期 2–4 月，长势良好。管护单位都江堰旅游管理局离堆公园管理处。

Located in the Bonsai Garden of the Lidui Park, Dujiangyan City, Sichuan, with latitude 30°59′52″ N, longitude 103°36′42″ E, elevation 686.0 m, tree H 4.1 m, CG 66.0 cm, CD 3.7 m/3.0 m, 110 y, Fl. Feb. to Apr. Grows well. Managed and maintained by Lidui Park Administrative Office of Dujiangyan Tourism Administration.

古4：氅盔
C. japonica 'Changkui'

位于四川省都江堰离堆公园入口左面，庭院地栽，N30°59′52″，E103°36′42″，海拔686.0 m。株高7.0 m，基围83.2 cm，冠幅6.2/5.7 m，树龄139年，花期2–4月，长势良好。管护单位都江堰旅游管理局离堆公园管理处。

Located at the left of the Lidui Park entrance, Dujiangyan city, Sichuan, with latitude 30°59′52″ N, longitude 103°36′42″ E, elevation 686.0 m, tree H 7.0 m, CG 83.2 cm, CD 6.2 m/5.7 m, 139 y, Fl. Feb. To Apr. Grows well. Managed and maintained by Lidui Park Administrative Office of Dujiangyan Tourism Administration.

雅安市及邛崃市

INTRODUCTION TO
YA'AN CITY AND QIONGLAI CITY

古 1：红绣球
C. japonica 'Hongxiuqiu'

位于四川省雅安市雨城区大里 8 组，N29°50′42″，E103°06′51″，海拔 817.5 m。高 3.7 m，基围 47.0 cm，冠幅 2.0 m，树龄 113 年。房前坎下洗衣池排水沟边，常年积水，叶片泛黄，有明显枯枝，长势差。为私家管护。

Located in Dali group 8 of Yucheng District, Ya'an City, Sichuan, with latitude 29°50′42″ N, longitude 103°06′51″ E, elevation 817.5 m, tree H 3.7 m, CG 47.0 cm, CD 2.0 m, 113 y. Planted near the drainage ditch of a laundry pool which has water accumulation all year round, that results yellow foliage and obvious dead twigs. Grows weakly. Private management.

古2：红绣球
C. japonica 'Hongxiuqiu'

位于四川省邛崃市文君街道东虹路191号附16号，N30°24′29″，E103°28′37″，海拔492.6 m。株高5.0 m，基围72.2 cm，冠幅4.3 m，树龄121年，花期1–4月，长势良好。据主人张绍伟讲述此株茶花来自四川蒲江成家镇，成家人清朝时弃官从商，从宫廷中流传出来，70岁老人从小看老，茶花无变化。私家管护。

Located at No. 191-16, Donghong road, Wenjun street, Qionglai City, Sichuan, with latitude 30°24′29″ N, longitude 103°28′37″ E, elevation 492.6 m, tree H 5.0 m, CG 72.2 cm, CD 4.3 m, 121 y, Fl. Jan. to Apr. Grows well. According to its owner Zhang Shaowei, it came from Chengjia town, Pujiang County, Sichuan. It was planted when the Cheng family abandoned the official business to start their own business in the Qing Dynasty. Private management.

古 3：氅盔
C. japonica 'Changkui'

位于四川省雅安市天全镇始阳街新村三组高术兵农家小院一角落，N30°01′26″，E102°49′02″，海拔 678.2 m。株高 5.6 m，基围 75.4 cm，冠幅 3.5 m，树龄 126 年，花期 1–4 月。树干基部因房屋改建和地坪处相连，已被砖和混凝土封住，生长环境恶劣，急需加强保护，长势良好。为私家管护。

Located at the Gao's yard of Xincun group 3 of Tianquan County, Ya'an City, Sichuan, with latitude 30°01′26″ N, longitude 102°49′02″ E, elevation 678.2 m, tree H 5.6 m, CG 75.4 cm, CD 3.5 m, 126 y, Fl. Jan. to Apr. The trunk base has been sealed with bricks and concrete. It lives in a harsh environment and needs protection urgently. Grows well. Private management.

古 4: 红绣球
C. japonica 'Hongxiuqiu'

位于四川省雅安市天全镇兴中村七组 11 号,高老师家,N30°01′13″,E102°49′48″,海拔 664.0 m。株高 6.5 m,基围 116.2 cm,冠幅 7.8 m,树龄 194 年,花期 2-4 月。生长在农家门前,地面是石材铺装,后期砌筑高于地面花台(约 70cm),树干基部因受花台限制,生长空间狭小,树冠离房屋太近,部分枝条已被人为损毁,急需加强保护,长势良好。私家管护。

Located in group 7 of Xingzhong Village of Tianquan County, Ya'an City, Sichuan, with latitude 30°01′13″ N, longitude 102°49′48″ E, elevation 664.0 m, tree H 6.5 m, CG 116.2 cm, CD 7.8 m, 194 y, Fl. Feb. to Apr. It grows in front of a farmer's house from brick ground, on where a ground flower platform was built later (around 70 cm). The trunk base is restricted by the flower platform and limited growth space. Its canopy is too close to the roof and some branches have been artificially damaged. It needs better protection urgently. Grows well. Private management.

古5：花洋红
C. japonica 'Hua Yanghong'

位于四川省雅安市天全镇兴中村七组11号，高老师家后院，N30°01′13″，E102°49′48″，海拔671.4 m。株高7.4 m，基围105.2 cm，冠幅5.0 m，树龄176年，花期1–4月。该茶花种植在高家农户院坝围墙边，据主人讲述是后来砌了一角花台，树干和根系透气性差，生长环境欠佳，树冠离房屋太近，已被人为损毁了近一半，急需要加强保护，长势良好。私家管护。

Located in group 7 of Xingzhong Village of Tianquan County, Ya'an City, Sichuan, with latitude 30°01′13″ N, longitude 102°49′48″ E, elevation 671.4 m, tree H 7.4 m, CG 105.2 cm, CD 5.0 m, 176 y, Fl. Jan. to Apr. It is planted at the corner of the backyard of farmer Gao's house. According to its owner, the air permeability of branches and roots is poor due to a later built flower platform. It lives in a harsh environment. Its canopy grows too close to roof and half of which has been artificially damaged. It needs better protection urgently. Grows well. Private management.

峨眉山景区
万年寺

INTRODUCTION TO WANNIAN TEMPLE IN MOUNT EMEI SCENIC AREA

古1：三学士
C. japonica 'Sanxueshi'

位于四川省峨眉山市峨眉山景区万年寺巍峨宝殿前花坛，N29°32′44″，E103°19′58″，海拔1020.0 m。株高4.8 m，基围62.0 cm，冠幅3.8 m，树龄102年，花期2–4月，长势良好。管护单位峨眉山风景名胜管理委员会。

Located at the flowerbed in front of the majestic palace of Wannian Temple in Mount Emei Scenic Area, Emeishan City, Sichuan, with latitude 29°32′44″ N, longitude 103°19′58″ E , elevation 1020.0 m, tree H 4.8 m, CG 62.0 cm, CD 3.8 m, 102 y, Fl. Feb. to Apr. Grows well. Managed and maintained by Mount Emei Scenic Area Management Committee.

彭州市
葛仙山风景区

***INTRODUCTION TO
GEXIAN MOUNTAIN SCENIC AREA,
PENGZHOU CITY***

古 1：川芙蓉
C. japonica 'Chuanfurong'

位于四川省彭州市葛仙山镇万年乡葛仙山风景区，N31°10′22″，E103°57′27″，海拔962.6 m。株高4.5 m，基围54.0 cm，冠幅4.2 m，树龄103年，花期2–4月，长势良好。管理单位彭州市葛仙山景区。

Located in the Gexian Mountain Scenic Area at Wannian Town, Pengzhou City, Sichuan, with latitude 31°10′22″ N, longitude 103°57′27″ E, elevation 962.6 m, tree H 4.5 m, CG 54.0 cm, CD 4.2 m, 103 y, Fl. Feb. to Apr. Grows well. Managed and maintained by Pengzhou City Gexian Mountain Scenic Area.

邛崃市
天台山风景区

**INTRODUCTION TO
QIONGLAI TIANTAI MOUNTAIN SCENIC AREA**

邛崃天台山风景区是四川省大熊猫栖息地世界自然遗产、国家重点风景名胜区、国家级森林公园、国家 AAAA 级旅游区。相传大禹率众治水，在此登高祭天，天台山由此得名。据不完全统计，景区内有植物 83 科 400 余种，种质资源丰富，约有上万株野生西南红山茶古树。

Qionglai Tiantai Mountain Scenic Area is one of the habitats for panda in Sichuan, while it is also a World Natural Heritage, National Key Scenic Spot, National Forest Park, and National AAAA Level Tourist Area. According to the legend, during the period of Yu led the crowd to control flood, he climbed to the mountain top to worship heaven, thus the mountain was named to Tiantai Mountain. According to incomplete statistics, there are more than 83 families and 400 species of plants in the scenic area with rich germplasm resources. There are about tens of thousands of wild ancient trees of *C. Pitardii* Coh. St. var. *Pitardii* Sealy.

古 1：西南白山茶
C. pitardii Coh. St. var. *alba* Chang

位于四川省邛崃天台山景区"知音泉"旁，N30°15′22″，E103°05′56″，海拔 1233.5 m。株高 8.5 m，三分枝 63.0/57.0/63.0 cm，冠幅 5.0 m，树龄 131 年。花期 11 月至翌年 3 月，长势好。管护单位四川省邛崃天台山国家级自然保护区。

Located at the Zhiyin Spring of Tiantai Mountain, Qionglai City, Sichuan, with latitude 30°15′22″ N, longitude 103°05′56″ E, elevation 1233.5 m, tree H 8.5 m, three branches (63.0 cm/57.0 cm/63.0 cm), CD 5.0 m, 131 y, Fl. Nov. to Mar. Grows well. Managed and maintained by Qionglai Tiantai Mountain National Nature Reserve.

古2：西南白山茶
C. pitardii Coh. St. var. *alba* Chang

位于四川省邛崃天台山景区路边，N30°15′23″，E103°05′22″，海拔 1262.6 m。株高 8.6 m，基围 151.0 cm，四分枝，最大三分枝基围 63.0/53.0/57.0 cm，冠幅 9.3m，树龄 235 年。花期 11 月至翌年 3 月。根茎裸露，抱石生长，长势好。管护单位四川省邛崃天台山国家级自然保护区。

Located at the roadside of Tiantai Mountain, Qionglai City, Sichuan, with latitude 30°15′23″ N, longitude 103°05′22″ E, elevation 1262.6 m, tree H 8.6 m, CG 151.0 cm, four branches (the biggest three 63.0 cm/53.0 cm/57.0 cm respectively), CD 9.3 m, 235 y, Fl. Nov. to Mar. Some roots are exposed and wrap stones. Grows well. Managed and maintained by Qionglai Tiantai Mountain National Nature Reserve.

古 3：西南白山茶
C. pitardii Coh. St. var. *alba* Chang

位于四川省邛崃天台山景区，N30°15′22″，E103°05′22″，海拔 1265.9 m。株高 9.7 m，基围 104.0 cm，冠幅 7.0 m，树龄 179 年。花期 11 月至翌年 3 月，长势良好。管护单位四川省邛崃天台山国家级自然保护区。

Located in the Tiantai Mountain of Qionglai City, Sichuan, with latitude 30°15′22″ N, longitude 103°05′22″ E, elevation 1265.9 m, tree H 9.7 m, CG 104.0 cm, CD 7.0 m, 179 y, Fl. Nov. to Mar. Grows well. Managed and maintained by Qionglai Tiantai Mountain National Nature Reserve.

古 4：西南白山茶
C. pitardii Coh. St.var. *alba* Chang

位于四川省邛崃天台山景区"同心石"景点游览道边，N30°15′20″，E103°05′44″，海拔 1240.9 m。株高 8.4 m，基围 96.0 cm，冠幅 7.6 m，树龄 170 年。花期 11 月至翌年 3 月。树冠向溪边倾斜。因景区建设道路铺装和溪流涨水冲刷，导致部分树根裸露，长势良好，需加强养护。管护单位四川省邛崃天台山国家级自然保护区。

Located at the Tongxinshi of Tiantai Mountain, Qionglai City, Sichuan, with latitude 30°15′20″ N, longitude 103°05′44″ E, elevation 1240.9 m, tree H 8.4 m, CG 96.0 cm, CD 7.6 m, 170 y, Fl. Nov. to Mar. Its canopy leans to the side of stream. Due to road construction and stream flushing, some roots are exposed. Grows well. It needs better protection and management. Managed and maintained by Qionglai Tiantai Mountain National Nature Reserve.

古5: 西南白山茶
C. pitardii Coh. St. var. *alba* Chang

位于四川省邛崃天台山景区，N30°15′22″，E103°05′46″，海拔1237.5 m。株高8.3 m，基围143.0 cm，六分枝，冠幅11.4 m，树龄226年。花期11月至翌年3月，其中一枝已枯死。管护单位四川省邛崃天台山国家级自然保护区。

Located in the Tiantai Mountain of Qionglai City, Sichuan, with latitude 30°15′22″ N, longitude 103°05′46″ E, elevation 1237.5 m, tree H 8.3 m, CG 143.0 cm, six branches, CD 11.4 m, 226 y, Fl. Nov. to Mar. One branch is dead. Managed and maintained by Qionglai Tiantai Mountain National Nature Reserve.

古6：西南白山茶
C. pitardii Coh. St. var. alba Chang

位于四川省邛崃天台山景区，N30°15′23″，E103°05′48″，海拔 1237.5 m。株高 7.6 m，基围 91.0 cm，冠幅 5.8 m，树龄 101 年。花期 11 月至翌年 3 月，长势良好。管护单位四川省邛崃天台山国家级自然保护区。

Located in the Tiantai Mountain of Qionglai City, Sichuan, with latitude 30°15′23″ N, longitude 103°05′48″ E, elevation 1237.5 m, tree H 7.6 m, CG 91.0 cm, CD 5.8 m, 101 y, Fl. Nov. to Mar. Grows well. Managed and maintained by Qionglai Tiantai Mountain National Nature Reserve.

古 7：西南白山茶
C. pitardii Coh. St. var. alba Chang

位于四川省邛崃天台山景区，N30°15′22″，E103°05′50″，海拔 1237.6 m。株高 18.0 m，基围 188.0 cm，四分枝 70.0/63.0/65.0/56.0 cm，冠幅 11.5 m，树龄 279 年。花期 11 月至翌年 3 月，长势良好。管护单位四川省邛崃天台山国家级自然保护区。

Located in the Tiantai Mountain of Qionglai City, Sichuan, with latitude 30°15′22″ N, longitude 103°05′50″ E, elevation 1237.6 m, tree H 18.0 m, CG 188.0 cm, four branches (70.0 cm/63.0 cm/65.0 cm/56.0 cm), CD 11.5 m, 279 y, Fl. Nov. to Mar. Grows well. Managed and maintained by Qionglai Tiantai Mountain National Nature Reserve.

古8：西南白山茶
C. pitardii Coh. St. var. *alba* Chang

位于四川省邛崃天台山景区，N30°15′22″，E103°05′52″，海拔 1235.7 m。株高 15.5 m，五分枝 55.0/47.0/47.0/61.0/59.0 cm，冠幅 7.7 m，树龄 122 年，花期 11 月至翌年 3 月，有一枝枯死，长势良好。管护单位四川省邛崃天台山国家级自然保护区。

Located in the Tiantai Mountain of Qionglai City, Sichuan, with latitude 30°15′22″ N, longitude 103°05′52″ E, elevation 1235.7 m, tree H 15.5 m, five branches (55.0 cm/47.0 cm/47.0 cm/61.0 cm/59.0 cm), CD 7.7 m, 122 y, Fl. Nov. to Mar. One branch is dead. Grows well. Managed and maintained by Qionglai Tiantai Mountain National Nature Reserve.

古 9：西南红山茶
C. pitardii Coh. St. var. *Pitardii* Sealy

位于四川省邛崃天台山景区"放翁亭"景点处，N30°15′58″，E103°06′45″，海拔 1200.6 m。株高 12.0 m，基围 141.0 cm，冠幅 7.0 m，树龄 170 年。花期 11 月至翌年 3 月。长势良好。管护单位四川省邛崃天台山国家级自然保护区。

Located at Fangweng Pavilion of Tiantai Mountain, Qionglai City, Sichuan, with latitude 30°15′58″ N, longitude 103°06′45″ E, elevation 1200.6 m, tree H 12.0 m, CG 141.0 cm, CD 7.0 m, 170 y, Fl. Nov. to Mar. Grows well. Managed and maintained by Qionglai Tiantai Mountain National Nature Reserve.

古10：西南白山茶
C. pitardii Coh. St. var. *alba* Chang

位于四川省邛崃天台山景区，N30°16′03″, E103°06′60″, 海拔1194.5 m。株高9.4 m，六分枝，最大两分枝94.0/100.0 cm，冠幅7.6 m，树龄176年。花期11月至翌年3月，长势好。管护单位四川省邛崃天台山国家级自然保护区。

Located in the Tiantai Mountain of Qionglai City, Sichuan, with latitude 30°16′03″ N, longitude 103°06′60″ E, elevation 1194.5 m, tree H 9.4 m, six branches (the biggest two 94.0 cm/100.0 cm respectively), CD 7.6 m, 176 y, Fl. Nov. to Mar. Grows well. Managed and maintained by Qionglai Tiantai Mountain National Nature Reserve.

古 11：西南白山茶
C. pitardii Coh. St. var. *alba* Chang

位于四川省邛崃天台山景区，N30°16′03″，E103°06′59″，海拔1194.4 m。株高8.9 m，七分枝，最大五分枝70.0/72.0/35.0/40.0/52.0 cm，冠幅7.4 m，树龄142年。花期11月至翌年3月，已死两枝。管护单位四川省邛崃天台山国家级自然保护区。

Located in the Tiantai Mountain of Qionglai City, Sichuan, with latitude 30°16′03″ N, longitude 103°06′59″ E, elevation 1194.4 m, tree H 8.9 m, seven branches (the biggest five 70.0 cm/72.0 cm/35.0 cm/40.0 cm/52.0 cm respectively), CD 7.4 m, 142 y, Fl. Nov. to Mar. Two branches are dead. Managed and maintained by Qionglai Tiantai Mountain National Nature Reserve.

古12：西南白山茶
C. pitardii Coh. St. var. *alba* Chang

位于四川省邛崃天台山景区，N30°16′11″，E103°07′01″，海拔1189.0 m。株高12.2 m，基围120.0 cm，冠幅10.2 m，树龄199年。花期11月至翌年3月，长势良好。管护单位四川省邛崃天台山国家级自然保护区。

Located in the Tiantai Mountain of Qionglai City, Sichuan, with latitude 30°16′11″ N, longitude 103°07′01″ E, elevation 1189.0 m, tree H 12.2 m, CG 120.0 cm, CD 10.2 m, 199 y, Fl. Nov. to Mar. Grows well. Managed and maintained by Qionglai Tiantai Mountain National Nature Reserve.

古 13: 西南红山茶
C. pitardii Coh. St. var. Pitardii Sealy

位于四川省邛崃天台山景区石林山坡上，N30°17′05″，E103°06′55″，海拔1184.2 m。株高6.6 m，基围117.0 cm，冠幅6.0 m，树龄145年，花期11月至翌年3月。主干已枯腐，现保留两分枝，一枝开白色花，一枝开粉色花，长势良好。管护单位四川省邛崃天台山国家级自然保护区。

Located at the backyard of Shilin in Tiantai Mountain, Qionglai City, Sichuan, with latitude 30°17′05″ N, longitude 103°06′55″ E, elevation 1184.2 m, tree H 6.6 m, CG 117.0 cm, CD 6.0 m, 145 y, Fl. Nov. to Mar. The main trunk has been rotted, two branches remain with white flowers and pink flowers respectively. Grows well. Managed and maintained by Qionglai Tiantai Mountain National Nature Reserve.

古 14：西南红山茶
C. pitardii Coh. St. var. *Pitardii* Sealy

位于四川省邛崃天台山景区趣园对面村民住宅旁，N30°17′13″，E103°07′03″，海拔 1182.3 m。株高 10.5 m，基围 100.0 cm，冠幅 6.7 m，树龄 120 年，花期 11 月至翌年 3 月，长势良好。管护单位四川省邛崃天台山国家级自然保护区。

Located at the villager's house opposite to the Quyuan of Tiantai Mountain, Qionglai City, Sichuan, with latitude 30°17′13″ N, longitude 103°07′03″ E, elevation 1182.3 m, tree H 10.5 m, CG 100.0 cm, CD 6.7m, 120 y, Fl. Nov. to Mar. Grows well. Managed and maintained by Qionglai Tiantai Mountain National Nature Reserve.

古 15：西南红山茶
C. pitardii Coh. St. var. *Pitardii* Sealy

位于四川省邛崃天台山景区趣园对面村民住宅旁，N30°17′13″，E103°07′03″，海拔 1183.4 m。株高 9.4 m，基围 107.0 cm，冠幅 7.2 m，树龄 115 年，花期 11 月至翌年 3 月。树干有空腐，虫蛀明显，需加强养护。管护单位四川省邛崃天台山国家级自然保护区。

Located at the villager's house opposite to the Quyuan of Tiantai Mountain, Qionglai City, Sichuan, with latitude 30°17′13″ N, longitude 103°07′03″ E, elevation 1183.4 m, tree H 9.4 m, CG 107.0 cm, CD 7.2 m, 115 y, Fl. Nov. to Mar. The trunk has been hollowly rotted, with obvious moth-eaten trace and need better protection and management. Managed and maintained by Qionglai Tiantai Mountain National Nature Reserve.

古 16: 西南红山茶
C. pitardii Coh. St. var. *Pitardii* Sealy

位于四川省邛崃天台山景区趣园对面村民住宅旁，N30°17′13″，E103°07′03″，海拔 1183.5 m。株高 9.6 m，基围 102.0 cm，冠幅 7.2 m，树龄 100 年，花期 11 月至翌年 3 月，长势良好。管护单位四川省邛崃天台山国家级自然保护区。

Located at the villager's house opposite to the Quyuan of Tiantai Mountain, Qionglai City, Sichuan, with latitude 30°17′13″ N, longitude 103°07′03″ E, elevation 1183.5 m, tree H 9.6 m, CG 102.0 cm, CD 7.2 m, 100 y, Fl. Nov. to Mar. Grows well. Managed and maintained by Qionglai Tiantai Mountain National Nature Reserve.

古17：西南红山茶
C. pitardii Coh. St. var. *Pitardii* Sealy

位于四川省邛崃市天台山自然保护区游览道边。N30°15′22″，E103°05′54″，海拔1235.0 m。株高12.0 m，六分枝，最大四分枝65.0/57.0/63.0/59.0 cm，冠幅11.0 m，树龄134年。花期11月至翌年3月。部分树根裸露，长势良好。管护单位四川省邛崃天台山国家级自然保护区。

Located at the roadside of Tiantai Mountain, Qionglai City, Sichuan, with latitude 30°15′22″ N, longitude 103°05′54″ E, elevation 1235.0 m, tree H 12.0 m, six branches (the biggest four 65.0 cm/57.0 cm/63.0 cm/59.0 cm respectively), CD 11.0 m, 134 y, Fl. Nov. to Mar. Some roots are exposed. Grows well. Managed and maintained by Qionglai Tiantai Mountain National Nature Reserve.

古 18：西南红山茶
C. pitardii Coh. St. var. Pitardii Sealy

位于四川省邛崃天台山景区小溪边，N30°15′21″，E103°05′41″，海拔1242.7 m。株高15.0 m，基围91.0 cm，冠幅6.0 m，树龄101年，花期11月至翌年3月。由于常年流水冲刷，部分树根裸露，树干基部空腐，长势良好，需加强养护。管护单位四川省邛崃天台山国家级自然保护区。

Located at the stream bank of Tiantai Mountain, Qionglai City, Sichuan, with latitude 30°15′21″ N, longitude 103°05′41″ E, elevation 1242.7 m, tree H 15.0 m, CG 91.0 cm, CD 6.0 m, 101 y, Fl. Nov. to Mar. Due to year-round stream flushing, some roots are exposed. The trunk base is hollowly rotted. Grows well. Need better protection and management. Managed and maintained by Qionglai Tiantai Mountain National Nature Reserve.

古 19：西南白山茶
C. pitardii Coh. St. var. *alba* Chang

位于四川省邛崃天台山景区"清风晓月"亭处，N30°15′21″，E103°05′42″，海拔1243.6 m。株高10.8 m，基围95.0 cm，冠幅7.0 m，树龄100年，花期11月至翌年3月，长势良好。管护单位四川省邛崃天台山国家级自然保护区。

Located at the Qingfengxiaoyue Pavilion in Tiantai Mountain, Qionglai City, Sichuan, with latitude 30°15′21″ N, longitude 103°05′42″ E, elevation 1243.6 m, tree H 10.8 m, CG 95.0 cm, CD 7.0 m, 100 y, Fl. Nov. to Mar. Grows well. Managed and maintained by Qionglai Tiantai Mountain National Nature Reserve.

古20：西南白山茶
C. pitardii Coh. St. var. *alba* Chang

位于四川省邛崃市天台山自然保护区主游览道边，N30°16′03″，E103°06′59″，海拔 1194.7 m。株高 12.5 m，三分枝，最大两分枝 91.0/79.0 cm，冠幅 7.7 m，树龄 165 年。花期 11 月至翌年 3 月，长势一般。管护单位四川省邛崃天台山国家级自然保护区。

Located at the main roadside of Tiantai Mountain, Qionglai City, Sichuan, with latitude 30°16′03″ N, longitude 103°06′59″ E, elevation 1194.7 m, tree H 12.5 m, three branches (the biggest two 91.0 cm/79.0 cm respectively), CD 7.7 m, 165 y, Fl. Nov. to Mar. Managed and maintained by Qionglai Tiantai Mountain National Nature Reserve.

古21：西南白山茶
C. pitardii Coh. St. var. *alba* Chang

位于四川省邛崃市天台山自然保护区主游览道边，N30°16'14″，E103°07'04″，海拔 1185.9 m。株高9.0 m，基围122.0 cm，冠幅8.0 m，树龄202年。花期11月至翌年3月，长势良好。管护单位四川省邛崃天台山国家级自然保护区。

Located at the main roadside of Tiantai Mountain, Qionglai City, Sichuan, with latitude 30°16'14″ N, longitude 103°07'04″ E, elevation 1185.9 m, tree H 9.0 m, CG 122.0 cm, CD 8.0 m, 202 y, Fl. Nov. to Mar. Grows well. Managed and maintained by Qionglai Tiantai Mountain National Nature Reserve.

古 22：西南红山茶
C. pitardii Coh. St. var. *Pitardii* Sealy

位于四川省邛崃市天台山自然保护区主游览道边。N30°16′14″，E103°07′04″，海拔 1186.1 m。株高 9.0 m，基围 104.0 cm，冠幅 8.5 m，树龄 179 年。花期 11 月至翌年 3 月，长势一般。管护单位四川省邛崃天台山国家级自然保护区。

Located at the main roadside of Tiantai Mountain, Qionglai City, Sichuan, with latitude 30°16′14″ N, longitude 103°07′04″ E, elevation 1186.1 m, tree H 9.0 m, CG 104.0 cm, CD 8.5 m, 179 y, Fl. Nov. to Mar. Managed and maintained by Qionglai Tiantai Mountain National Nature Reserve.

古 23：西南红山茶
C. pitardii Coh. St. var. *Pitardii* Sealy

位于四川省邛崃天台山景区小溪边，N30°15′21″，E103°05′41″，海拔 1245.5 m。株高 8.0 m，基围 110.0 cm，三分枝，冠幅 7.5 m，树龄 130 年。花期 11 月至翌年 3 月，主分枝已枯死一枝，长势良好。管护单位四川省邛崃天台山国家级自然保护区。

Located at the stream bank of Tiantai Mountain, Qionglai City, Sichuan, with latitude 30°15′21″ N, longitude 103°05′41″ E, elevation 1245.5 m, tree H 8.0 m, CG 110.0 cm, three branches, CD 7.5 m, 130 y, Fl. Nov. to Mar. One of the main branches is dead. Grows well. Managed and maintained by Qionglai Tiantai Mountain National Nature Reserve.

INTRODUCTION TO
EMEISHAN CITY

峨眉山位于四川省西南部，生物种类丰富，拥有高等植物242科3200多种，是一个集自然风光与佛教文化于一体的国家级风景名胜区，在峨眉山洪雅县周边分布有大量野生峨眉红山茶古树。

据当地村民介绍，2000年前后从峨眉山周边山上移走一大批峨眉红山茶大树，目前原生境茶花大树已经很少见了，现有峨眉红山茶均为2005年后从四川省眉山市洪雅县周边山上移栽过来。

Located in the southwest of Sichuan, Mount Emei is rich in biological species and has more than 3,200 species of 242 families of higher plants. It is a national-level scenic spot integrating natural scenery and Buddhist culture. There are lots of wild *C. omeiensis* Chang distributed in Hongya County, Emeishan City.

According to local people, a large number of *C. omeiensis* Chang trees were removed from surrounding mountains of Mount Emei around 2000. Now there are few native ancient camellia trees. After 2005, the current ones were transplanted from the surrounding mountains of Hongya County, Emeishan City, Sichuan.

古 1：峨眉红山茶
C. omeiensis Chang

位于四川省雅安市雨城区大里 8 组，N29°50′42″，E103°06′51″，海拔 824.5 m。高 3.8 m，基围 95.0 cm，冠幅 2.3 m，树龄 170 年。菜园地栽种，长势良好，为私家管护。

Located in Dali group 8 of Yucheng District, Ya'an City, Sichuan, with latitude 29°50′42″ N, longitude 103°06′51″ E, elevation 824.5 m, tree H 3.8 m, CG 95.0 cm, CD 2.3 m, 170 y. Planted in the vegetable field. Grows well. Private management.

古2：峨眉红山茶
C. omeiensis Chang

位于四川省雅安市雨城区大里 8 组，N29°50′42″，E103°06′51″，海拔 824.4 m。高 3.9 m，基围 80.0 cm，冠幅 2.2 m，树龄 152 年。菜园地栽种，长势良好，为私家管护。

Located in Dali group 8 of Yucheng District, Ya'an City, Sichuan, with latitude 29°50′42″ N, longitude 103°06′51″ E, elevation 824.4 m, tree H 3.9 m, CG 80.0 cm, CD 2.2 m, 152 y. Planted in the vegetable field. Grows well. Private management.

古 3：峨眉红山茶
C. omeiensis Chang

位于四川省雅安市雨城区大里 8 组，N29°50′42″，E103°06′51″，海拔 824.1 m。高 3.3 m，基围 65.0 cm，冠幅 3.2 m，树龄 134 年。菜园地栽种，长势良好，为私家管护。

Located in Dali group 8 of Yucheng District, Ya'an City, Sichuan, with latitude 29°50′42″ N, longitude 103°06′51″ E, elevation 824.1 m, tree H 3.3 m, CG 65.0 cm, CD 3.2 m, 134 y. Planted in the vegetable field. Grows well. Private management.

古 4：峨眉红山茶
C. omeiensis Chang

位于四川省雅安市雨城区大里 8 组，N29°50′43″，E103°06′52″，海拔 836.1 m。高 3.4 m，基围 110.0 cm，冠幅 3.0 m，树龄 187 年。主干空腐，房屋后坡地栽种，长势良好，为私家管护。

Located in Dali group 8 of Yucheng District, Ya'an City, Sichuan, with latitude 29°50′43″ N, longitude 103°06′52″ E, elevation 836.1 m, tree H 3.4 m, CG 110.0 cm, CD 3.0 m, 187 y. The main trunk is hollowly rotted. Planted in the hillside field behind the house. Grows well. Private management.

古 5: 峨眉红山茶
C. omeiensis Chang

位于四川省雅安市雨城区大里 8 组，N29°50′43″，E103°06′52″，海拔 836.5 m。高 3.4 m，基围 80.0 cm，冠幅 3.0 m，树龄 152 年。房屋后坡地栽种，长势良好，为私家管护。

Located in Dali group 8 of Yucheng District, Ya'an City, Sichuan, with latitude 29°50′43″ N, longitude 103°06′52″ E, elevation 836.5 m, tree H 3.4 m, CG 80.0 cm, CD 3.0 m, 152 y. Planted in the hillside field behind the house. Grows well. Private management.

古6：峨眉红山茶
C. omeiensis Chang

位于四川省雅安市雨城区大里 8 组，N29°50′43″，E103°06′52″，海拔 836.9 m。高 4.9 m，基围 63.0 cm，冠幅 3.3 m，树龄 132 年。主干空腐，只剩半张树皮，房屋后坡地栽种，长势良好，为私家管护。

Located in Dali group 8 of Yucheng District, Ya'an City, Sichuan, with latitude 29°50′43″ N, longitude 103°06′52″ E, elevation 836.9 m, tree H 4.9 m, CG 63.0 cm, CD 3.3 m, 132 y. The main trunk is hollowly rotted. Planted in the hillside field behind the house. Grows well. Private management.

古 7：峨眉红山茶
C. omeiensis Chang

位于四川省雅安市雨城区许桥村 3 组，N29°47′59″，E103°06′56″，海拔 769.4m。高 3.2 m，基围 60.0 cm，冠幅 2.1 m，树龄 128 年。菜园地栽种，长势良好，为私家管护。

Located in group 3 of Xuqiao Village of Yucheng District, Ya'an City, Sichuan with latitude 29°47′59″ N, longitude 103°06′56″ E, elevation 769.4 m, tree H 3.2 m, CG 60.0 cm, CD 2.1 m, 128 y. Planted in the vegetable field. Grows well. Private management.

古 8：峨眉红山茶
C. omeiensis Chang

位于四川省雅安市雨城区许桥村 3 组，N29°47′59″，E103°06′56″，海拔 769.3 m，高 3.0 m，基围 60.0 cm，冠幅 1.7 m，树龄 128 年。菜园地栽种，长势良好，为私家管护。

Located in group 3 of Xuqiao Village of Yucheng District, Ya'an City, Sichuan with latitude 29°47′59″ N, longitude 103°06′56″ E, elevation 769.3 m, tree H 3.0 m, CG 60.0 cm, CD 1.7 m, 128 y. Planted in the vegetable field. Grows well. Private management.

附表 部分古树及后备古树基础资料

Table-basic data of some ancient trees(aged over 100) and trees (aged 80-100)

序号 NO.	种/品种 Cultivar	高度(m) Height	冠幅(m) CD	基围(cm) CG	最大三枝周长 The biggest three circumference	树龄(年) Year	海拔(m) Elevation	纬度 Latitude	经度 Longitude	生长地点 Location
1	'胭脂鳞'	3.1	3.5	46.0	38/23.5	101	540.7	N29°33′27″	E106°37′19″	重庆市南山植物园
2	'白宝塔'	4.5	4.8	46.0	41/28	104	541.1	N29°33′27″	E106°37′18″	重庆市南山植物园
3	'紫金冠'	5.3	4.6	46.0	31/35	112	541.2	N29°33′27″	E106°37′18″	重庆市南山植物园
4	'紫金冠'	4.8	4.3	65.0	44/39	112	541.0	N29°33′27″	E106°37′18″	重庆市南山植物园
5	'白宝塔'	4.0	4.0		29/30	104	541.1	N29°33′26″	E 106°37′18″	重庆市南山植物园
6	'川牡丹茶'	3.6	4.0	49.0	30/33	112	541.1	N29°33′26″	E106°37′18″	重庆市南山植物园
7	'白洋片'	3.8	3.5	47.0		106	541.1	N29°33′26″	E106°37′18″	重庆市南山植物园
8	'七心红'	3.5	3.6	49.0	34.5/28.5	113	541.0	N29°33′26″	E106°37′18″	重庆市南山植物园
9	'花洋红'	4.0	4.0	48.5	31/43	123	541.0	N29°33′26″	E106°37′19″	重庆市南山植物园
10	'花洋红'	3.7	4.0	46.0	33/25	107	541.1	N29°33′26″	E106°37′19″	重庆市南山植物园
11	'七心大红'	4.0	4.2	35.0		116	539.8	N29°33′28″	E106°37′21″	重庆市南山植物园
12	'三学士'	4.8	3.5	33.5		104	539.1	N29°33′28″	E106°37′21″	重庆市南山植物园
13	'白洋片'	5.3	4.8		29.5/24.5	106	538.8	N29°33′28″	E106°37′21″	重庆市南山植物园
14	'白洋片'	5.0	5.2	53.0	28.5/29.5/30	106	538.0	N29°33′28″	E106°37′21″	重庆市南山植物园
15	'三学士'	4.5	5.0	44.5	33.5/30.5	104	539.5	N29°33′28″	E106°37′22″	重庆市南山植物园
16	'白洋片'	5.0	5.2	58.5	34/41.5	106	536.3	N29°33′28″	E106°37′23″	重庆市南山植物园
17	'胭脂鳞'	5.0	4.2	45.5		101	536.1	N29°33′28″	E 106°37′23″	重庆市南山植物园
18	'七心红'	2.0	1.8	35.0		113	535.4	N29°33′29″	E106°37′23″	重庆市南山植物园
19	'花五宝'	3.0	2.5	39.5		107	535.2	N29°33′29″	E106°37′23″	重庆市南山植物园
20	'青心白'	3.8	4.0	34.5		104	534.4	N29°33′29″	E106°37′23″	重庆市南山植物园
21	'白洋片'	3.6	2.5	38.5		106	533.7	N29°33′29″	E106°37′24″	重庆市南山植物园
22	'花洋红'	5.0	3.8	45.0		107	533.1	N29°33′29″	E106°37′24″	重庆市南山植物园
23	'三学士'	4.8	3.5		29.5/24	104	532.3	N29°33′29″	E106°37′24″	重庆市南山植物园

续表

序号	种/品种	高度(m)	冠幅(m)	基围(cm)	最大三枝周长	树龄(年)	海拔(m)	纬度	经度	生长地点
NO.	Cultivar	Height	CD	CG	The biggest three circumference	Year	Elevation	Latitude	Longitude	Location
24	'白宝塔'	4.5	3.5	46.0		139	530.2	N29°33'29"	E106°37'24"	重庆市南山植物园
25	'三学士'	4.0	4.0	47.0		138	529.5	N29°33'29"	E106°37'24"	重庆市南山植物园
26	'花洋红'	3.5	4.0	44.0	25.5/29	103	540.9	N29°33'26"	E106°37'19"	重庆市南山植物园
27	'川牡丹茶'	4.0	3.5	46.5	37	103	531.7	N29°33'26"	E106°37'19"	重庆市南山植物园
28	'花洋红'	4.2	3.5	46.0	33.5	107	547.1	N29°33'26"	E106°37'19"	重庆市南山植物园
29	'白宝塔'	3.5	4.2	47.0		139	540.9	N29°33'26"	E106°37'19"	重庆市南山植物园
30	'七心红'	3.0	3.0	44.0		113	540.9	N29°33'26"	E106°37'19"	重庆市南山植物园
31	'小红莲'	3.7	3.2	39.0		104	540.9	N29°33'26"	E106°37'19"	重庆市南山植物园
32	'七心红'	3.5	5.3	45.0	31/24.5	113	540.9	N29°33'26"	E106°37'19"	重庆市南山植物园
33	'花洋红'	4.2	3.5	49.0	28/27	123	541.0	N29°33'26"	E106°37'19"	重庆市南山植物园
34	'七心红'	3.0	4.0	47.0		113	541.0	N29°33'26"	E106°37'19"	重庆市南山植物园
35	'胭脂鳞'	1.8	1.8	38.0		101	541.5	N29°33'26"	E106°37'19"	重庆市南山植物园
36	'胭脂鳞'	1.2	1.2	29.0		101	541.4	N29°33'26"	E106°37'19"	重庆市南山植物园
37	'金顶大红'	1.5	1.8			139	541.0	N29°33'26"	E106°37'20"	重庆市南山植物园
38	'花洋红'	3.8	3.8	44.0	33.5	107	540.2	N29°33'26"	E106°37'20"	重庆市南山植物园
39	'重庆红'	3.8	3.0		31.5/18	106	539.2	N29°33'26"	E 106°37'20"	重庆市南山植物园
40	'白洋片'	3.8	4.0	44.0	31/29	104	538.9	N29°33'26"	E106°37'20"	重庆市南山植物园
41	'醉杨妃'	3.0	3.4	37.5		112	538.8	N29°33'26"	E106°37'20"	重庆市南山植物园
42	'川牡丹茶'	3.2	3.5	46.0	38/22.5	112	536.7	N29°33'26"	E106°37'20"	重庆市南山植物园
43	'白洋片'	4.3	4.6	46.0	27.5/32	106	536.5	N29°33'26"	E106°37'20"	重庆市南山植物园
44	'花洋红'	4.8	4.5	47.0	33/27.5	107	537.4	N29°33'26"	E106°37'19"	重庆市南山植物园
45	'三学士'	4.2	3.0	48.0	31/27.5	104	537.2	N29°33'26"	E106°37'20"	重庆市南山植物园
46	'白洋片'	3.1	3.0	43.0	26/28.5	106	537.0	N29°33'26"	E106°37'19"	重庆市南山植物园
47	西南红山茶	5.5	4.0	57.0		113	536.5	N29°33'26"	E106°37'19"	重庆市南山植物园
48	'白洋片'	4.0	4.3	43.3		106	535.6	N29°33'26"	E106°37'20"	重庆市南山植物园
49	'胭脂鳞'	3.5	3.8	39.9		101	535.6	N29°33'26"	N29°33'26"	重庆市南山植物园

续表

序号	种/品种	高度(m)	冠幅(m)	基围(cm)	最大三枝周长	树龄(年)	海拔(m)	纬度	经度	生长地点
NO.	Cultivar	Height	CD	CG	The biggest three circumference	Year	Elevation	Latitude	Longitude	Location
50	'胭脂鳞'	3.0	3.6	38.3		101	534.9	N29°33′26″	E106°37′20″	重庆市南山植物园
51	'三学士'	4.8	3.2	41.8		104	535.5	N29°33′26″	E106°37′20″	重庆市南山植物园
52	'白洋片'	3.5	3.3	37.1		106	535.6	N29°33′26″	E106°37′20″	重庆市南山植物园
53	'白洋片'	5.0	3.8	42.4		106	535.7	N29°33′25″	E106°37′20″	重庆市南山植物园
54	'白宝塔'	2.5	3.5	38.0		139	527.3	N29°33′25″	E106°37′20″	重庆市南山植物园
55	'醉杨妃'	3.5	3.4	44.0		112	525.6	N29°33′25″	E106°37′21″	重庆市南山植物园
56	'花洋红'	4.0	3.1	48.4		123	521.6	N29°33′25″	E106°37′22″	重庆市南山植物园
57	'金顶大红'	2.6	3.9	42.1		113	524.6	N29°33′25″	E106°37′22″	重庆市南山植物园
58	'肥鳖茶'	2.5	3.0	37.4		104	524.5	N29°33′25″	E106°37′22″	重庆市南山植物园
59	'川玛瑙'	3.5	3.2	39.3		101	514.0	N29°33′24″	E106°37′23″	重庆市南山植物园
60	'花洋红'	3.3	4.4	40.8		107	525.3	N29°33′26″	E106°37′22″	重庆市南山植物园
61	'重庆红'	2.5	3.6	40.5		149	525.9	N29°33′26″	E106°37′22″	重庆市南山植物园
62	'红佛鼎'	2.8	3.3	43.6		103	531.5	N29°33′26″	E106°37′21″	重庆市南山植物园
63	'醉杨妃'	3.0	3.7	43.0		112	532.1	N29°33′26″	E106°37′21″	重庆市南山植物园
64	'紫金冠'	7.5	5.3	39.3		152	537.4	N29°33′29″	E106°37′23″	重庆市南山植物园
65	'金顶大红'	3.3	4.5	41.8		113	523.5	N29°33′30″	E106°37′25″	重庆市南山植物园
66	'胭脂鳞'	2.5	4.7		38.9/35.2	101	459.2	N29°33′28″	E106°37′43″	重庆市南山植物园
67	'七心红'	2.5	4.4	47.7		113	460.8	N29°33′28″	E106°37′44″	重庆市南山植物园
68	'紫金冠'	3.6	4.4		35.2/29.8	112	460.6	N29°33′29″	E106°37′44″	重庆市南山植物园
69	'花五宝'	3.5	4.3	38.9		101	460.1	N29°33′29″	E106°37′44″	重庆市南山植物园
70	'胭脂鳞'	3.0	3.1	37.1		101	461.2	N29°33′29″	E 106°37′43″	重庆市南山植物园
71	'紫金冠'	3.5	3.7	37.1		101	461.1	N29°33′30″	E106°37′44″	重庆市南山植物园
72	'七心白'	2.8	2.6	38.0		106	461.2	N29°33′30″	E106°37′44″	重庆市南山植物园
73	'胭脂鳞'	3.8	3.3	36.1		101	460.6	N29°33′30″	E106°37′44″	重庆市南山植物园
74	'白洋片'	3.0	3.2	35.8		101	460.5	N29°33′30″	E106°37′44″	重庆市南山植物园
75	'醉杨妃'	2.5	2.4	27.6		101	460.7	N29°33′30″	E106°37′44″	重庆市南山植物园

续表

序号 NO.	种/品种 Cultivar	高度(m) Height	冠幅(m) CD	基围(cm) CG	最大三枝周长 The biggest three circumference	树龄(年) Year	海拔(m) Elevation	纬度 Latitude	经度 Longitude	生长地点 Location
76	'花丝莲'	4.0	2.5	40.2		104	460.5	N29°33′30″	E106°37′44″	重庆市南山植物园
77	'醉杨妃'	3.0	3.3	44.0		101	459.8	N29°33′30″	E106°37′44″	重庆市南山植物园
78	'胭脂鳞'	2.8	2.2	45.2		101	460.9	N29°33′30″	E106°37′44″	重庆市南山植物园
79	'醉杨妃'	3.0	2.6	46.5		112	461.4	N29°33′30″	E106°37′44″	重庆市南山植物园
80	'九心十八瓣'	5.0	3.5	49.6		139	526.6	N29°33′26″	E106°37′22″	重庆市南山植物园
81	'三学士'	3.0	3.3	32.3		104	528.1	N29°33′25″	E106°37′21″	重庆市南山植物园
82	'胭脂鳞'	3.5	4.1		32/27.9/22.3	101	522.9	N29°33′30″	E106°37′25″	重庆市南山植物园
83	'花洋红'	3.5	3.5	33.0		107	570.1	N29°33′26″	E106°37′21″	重庆市南山植物园
84	'白洋片'	4.5	4.0	41.0		106	564.3	N29°33′26″	E106°37′20″	重庆市南山植物园
85	'花洋红'	4.6	4.0	42.0		107	569.6	N29°33′26″	E106°37′20″	重庆市南山植物园
86	'九心十八瓣'	5.6	4.5	48.0		110	554.0	N29°33′26″	E106°37′20″	重庆市南山植物园
87	'花洋红'	4.2	4.5		26/28	101	526.1	N29°33′25″	E106°37′22″	重庆市南山植物园
88	'三学士'	3.6	3.5		22/24	101	522.2	N29°33′25″	E106°37′22″	重庆市南山植物园
89	'紫金冠'	6.2	3.0	42.0		112	521.8	N29°33′25″	E106°37′22″	重庆市南山植物园
90	'紫金冠'	2.5	2.5	46.0		112	580.8	N29°33′28″	E106°37′19″	重庆市南山植物园
91	'七心红'	1.6	2.7	36.8		140	186.5	N30°48′18″	E108°22′52″	万州西山公园
92	'七心红'	2.4	3.7	47.1		100	185.3	N30°48′19″	E108°22′53″	万州西山公园
93	'洋红'	2.8	5.1	45.2		91	185.2	N30°48′20″	E108°22′52″	万州西山公园
94	'洋红'	4.2	4.7	58.1		129	186.4	N30°48′17″	E108°22′50″	万州西山公园
95	'紫金冠'	4.9	4.9	59.9		133	183.7	N30°48′16″	E108°22′49″	万州西山公园
96	'七心红'	2.5	3.7	46.7		100	182.5	N30°48′16″	E108°22′51″	万州西山公园
97	'白宝塔'	5.0	4.9	43.9		93	182.8	N30°48′18″	E108°22′53″	万州西山公园
98	'紫金冠'	4.3	6.0	79.1		181	179.8	N30°48′18″	E108°22′53″	万州西山公园
99	'七心红'	2.4	3.7	55.2		121	179.6	N30°48′16″	E108°22′50″	万州西山公园
100	'洋红'	4.2	4.6	60.4		133	181.9	N30°48′17″	E108°22′43″	万州西山公园
101	'洋红'	2.8	4.6	45.2		93	180.5	N30°48′19″	E108°22′51″	万州西山公园

续表

序号	种/品种	高度(m)	冠幅(m)	基围(cm)	最大三枝周长	树龄(年)	海拔(m)	纬度	经度	生长地点
NO.	Cultivar	Height	CD	CG	The biggest three circumference	Year	Elevation	Latitude	Longitude	Location
102	'胭脂鳞'	5.0	5.0	47.6		100	179.7	N30°48′19″	E108°22′51″	万州西山公园
103	'胭脂鳞'	3.6	4.5	47.8		100	180.3	N30°48′17″	E108°22′52″	万州西山公园
104	'胭脂鳞'	4.7	4.5	47.8		100	181.2	N30°48′18″	E108°22′51″	万州西山公园
105	'洋红'	3.2	4.6	49.3		106	181.5	N30°48′17″	E108°22′52″	万州西山公园
106	'胭脂鳞'	2.5	3.0	18.5		100	181.7	N30°48′19″	E108°22′52″	万州西山公园
107	'洋红'	2.0	2.6	48.5		102	182.3	N30°48′18″	E108°22′53″	万州西山公园
108	'洋红'	3.7	4.5	44.5		93	183.6	N30°48′17″	E108°22′49″	万州西山公园
109	'紫金冠'	3.6	5.1	55.2		121	182.8	N30°48′18″	E108°22′51″	万州西山公园
110	闽鄂山茶	11.1	7.5	60.2		133	198.5	N30°48′19″	E108°22′51″	万州西山公园
111	'洋红'	2.9	2.7	35.4		101	187.3	N30°48′15″	E108°22′51″	万州西山公园
112	'胭脂鳞'	2.9	3.7	35.8		102	189.9	N30°48′18″	E108°22′51″	万州西山公园
113	'洋红'	2.7	2.7	39.8		102	195.9	N30°48′19″	E108°22′50″	万州西山公园
114	'胭脂鳞'	2.2	2.7	21.4		100	190.1	N30°48′19″	E108°22′52″	万州西山公园
115	'七心红玛瑙'	3.2	3.2	31.4		123	192.9	N30°48′16″	E108°22′48″	万州西山公园
116	'胭脂鳞'	2.9	3.2	30.8		110	192.9	N30°48′17″	E108°22′49″	万州西山公园
117	'紫金冠'	4.3	2.8	31.3		100	182.0	N29°41′10″	E105°43′00″	万州西山公园
118	'七心红玛瑙'	2.5	2.7	38.9		104	192.7	N30°48′17″	E108°22′49″	万州西山公园
119	'七心红'	2.3	3.7	34.5		107	191.7	N30°48′17″	E108°22′49″	万州西山公园
120	'洋红'	3.1	3.7	32.9		105	187.9	N30°48′15″	E108°22′49″	万州西山公园
121	'洋红'	3.1	4.2	38.9		123	187.8	N30°48′17″	E108°22′52″	万州西山公园
122	'黑艳红'	2.2	2.2	29.1		100	193.2	N30°48′18″	E108°22′51″	万州西山公园
123	'洋红'	3.6	4.5	38.9		134	186.6	N30°48′17″	E108°22′52″	万州西山公园
124	'胭脂鳞'	4.5	4.2	33.9		121	187.7	N30°48′17″	E108°22′52″	万州西山公园
125	'胭脂鳞'	4.1	4.2	35.2		107	188.0	N30°48′18″	E108°22′52″	万州西山公园
126	'胭脂鳞'	4.0	4.2	41.2		142	187.8	N30°48′18″	E108°22′52″	万州西山公园
127	'紫金冠'	3.6	3.3	38.5		156	187.3	N30°48′18″	E108°22′52″	万州西山公园

续表

序号	种/品种	高度(m)	冠幅(m)	基围(cm)	最大三枝周长	树龄(年)	海拔(m)	纬度	经度	生长地点
NO.	Cultivar	Height	CD	CG	The biggest three circumference	Year	Elevation	Latitude	Longitude	Location
128	'洋红'	3.6	3.5	41.6		197	187.2	N30°48'18"	E108°22'52"	万州西山公园
129	'洋红'	3.7	4.0	42.8		210	188.3	N30°48'18"	E108°22'52"	万州西山公园
130	'七心红'	2.6	2.2	42.1		108	189.8	N30°48'18"	E108°22'51"	万州西山公园
131	'胭脂鳞'	3.2	3.2	33.6		151	189.1	N30°48'17"	E108°22'52"	万州西山公园
132	'胭脂鳞'	4.3	3.7	186.9		100	186.9	N30°48'18"	E108°22'52"	万州西山公园
133	'胭脂鳞'	4.2	3.9	42.0		108	187.3	N30°48'18"	E108°22'52"	万州西山公园
134	'洋红'	3.6	4.5	41.8		117	187.5	N30°48'18"	E108°22'52"	万州西山公园
135	'七心红'	2.6	3.7	40.8		104	191.5	N30°48'18"	E108°22'51"	万州西山公园
136	'七心红'	2.6	2.7	31.4		100	193.6	N30°48'22"	E108°22'52"	万州西山公园
137	'新品种'	2.6	3.2	33.8		131	195.1	N30°48'21"	E108°22'52"	万州西山公园
138	'紫金冠'	2.8	2.8	31.3		107	196.5	N30°48'21"	E108°22'52"	万州西山公园
139	'金盘荔枝'	3.6	3.8	38.5		112	194.4	N30°48'21"	E108°22'52"	万州西山公园
140	'白宝塔'	1.8	3.4	35.9		101	195.4	N30°48'21"	E108°22'52"	万州西山公园
141	'金盘荔枝'	2.6	3.3	32.5		100	194.7	N30°48'21"	E108°22'52"	万州西山公园
142	'洋红'	4.1	3.2	41.4		102	188.4	N30°48'18"	E108°22'52"	万州西山公园
143	'胭脂鳞'	3.2	2.6	18.2		100	174.3	N30°48'22"	E108°22'57"	万州西山公园
144	'白宝塔'	2.3	2.4	31.9		101	193.5	N30°48'19"	E108°22'51"	万州西山公园
145	'七心红玛瑙'	3.2	2.7	29.8		100	174.2	N30°48'22"	E108°22'57"	万州西山公园
146	'胭脂鳞'	3.2	3.3	33.3		106	173.2	N30°48'21"	E108°22'57"	万州西山公园
147	'红绣球'	3.6	3.7	38.4		108	171.3	N30°48'21"	E108°22'58"	万州西山公园
148	'红绣球'	3.6	3.7	36.1		100	172.1	N30°48'22"	E108°22'57"	万州西山公园
149	'洋红'	4.7	4.7	38.8		167	188.5	N30°48'18"	E108°22'52"	万州西山公园
150	'胭脂鳞'	2.8	2.7	42.0		129	191.4	N30°48'18"	E108°22'51"	万州西山公园
151	西南红山茶	4.0	1.1	23.0		83	1319.6	N30°36'43"	E108°45'02"	万州区恒合乡黄桐寨
152	'白宝塔'	2.7	3.1	41.5		88	207.4	N29°49'02"	E106°24'53"	西南大学
153	'小红莲'	2.8	3.3	44.0		94	204.8	N29°49'01"	E106°24'53"	西南大学

续表

序号	种/品种	高度(m)	冠幅(m)	基围(cm)	最大三枝周长	树龄(年)	海拔(m)	纬度	经度	生长地点
NO.	Cultivar	Height	CD	CG	The biggest three circumference	Year	Elevation	Latitude	Longitude	Location
154	'紫金冠'	3.4	2.5	38.0		80	204.2	N 29°49′01″	E 106°24′53″	西南大学
155	'小红莲'	2.6	2.6	38.0		80	204.4	N 29°49′01″	E 106°24′53″	西南大学
156	'白宝塔'	3.0	2.7	40.0		84	204.6	N29°49′00″	E 106°24′52″	西南大学
157	'七心红'	3.0	2.8	45.0		96	205.9	N29°48′60″	E 106°24′53″	西南大学
158	'白宝塔'	3.1	2.8	44.0		94	204.7	N29°49′02″	E 106°24′53″	西南大学
159	'醉杨妃'	2.7	2.4	44.0		94	206.4	N29°48′60″	E 106°24′52″	西南大学
160	'白宝塔'	4.2	3.6	40.0		103	244.7	N29°33′13″	E 106°27′07″	沙坪公园
161	'醉杨妃'	4.8	3.8	42.0		102	245.0	N29°33′14″	E106°27′07″	沙坪公园
162	'金盘荔枝'	3.1	2.7	46.0		93	506.7	N 29°34′14″	E106°36′55″	重庆抗战遗址博物馆
163	'三学士'	3.0	1.9	47.0		96	504.1	N 29°34′14″	E106°36′54″	重庆抗战遗址博物馆
164	'金盘荔枝'	4.2	3.6	45.0		91	509.5	N 29°34′10″	E106°36′55″	重庆抗战遗址博物馆
165	'白洋片'	5.0	3.3	41.0		80	502.1	N 29°34′10″	E106°36′54″	重庆抗战遗址博物馆
166	'七心红'	3.5	2.7	45.0		96	181.6	N29°49′51″	E106°26′17″	北碚公园
167	'七心红'	3.0	3.7	45.0		96	179.0	N29°49′51″	E106°26′17″	北碚公园
168	'胭脂鳞'	4.5	3.5	45.0		91	267.1	N 29°02′32″	E 106°08′13″	重庆市聚奎中学校
169	'胭脂鳞'	3.9	3.3	43.0		85	269.9	N 29°02′32″	E 106°08′13″	重庆市聚奎中学校
170	'红绣球'	4.6	2.8	52.0		83	1131.1	N31°01′44″	E103°36′44″	都江堰灵岩寺
171	'白洋片'	4.0	2.7	48.0		80	671.9	N30°59′52″	E103°36′44″	四川省都江堰离堆公园
172	'洋红'	3.9	3.4/2.7	49.0		82	668.0	N30°59′52″	E103°36′42″	四川省都江堰离堆公园
173	西南红山茶	5.0	2.6/2.3	69.0		139	1222.8	N30°15′37″	E103°06′02″	四川省邛崃市天台山景区
174	'早白阳'	4.8	2.3/1.8	46.0		82	995.4	N31°10′13″	E103°57′20″	四川省彭州市葛仙山风景区
175	'白玉片'	3.5	2.5/2.3	45.0		83	940.6	N31°10′26″	E103°57′19″	四川省彭州市葛仙山风景区
176	'小红莲'	3.5	3.5/3.8	44.0		80	952.6	N31°10′29″	E103°57′17″	四川省彭州市葛仙山风景区
177	'三学士'	3.0	2.5/2.8	45.0		83	916.5	N31°10′25″	E103°57′18″	四川省彭州市葛仙山风景区

注：CD 冠幅；CG 基围。

参考文献

[1] 高继银，PARKS C R，杜跃强. 山茶属植物主要原种彩色图集 [M]. 杭州：浙江科学技术出版社，2005.

[2] 管开云，李纪元，王仲朗. 中国茶花图鉴 [M]. 杭州：浙江科学技术出版社，2014.

[3] LI J Y, NI S, WU H, et al. A rapid non-destructive tool to measure ages of living camellias[A]. Proceedings of Conghua International Historic Camellia Conservation Meeting[C], 2019:23-30.

[4] LY/T 2814—2017，川山茶栽培技术规程 [S].

[5] 王雨茜，高露双，赵秀海. 长白山山杨年轮年表及其与气候变化的关系 [J]. 东北林业大学学报，2013，41(1)：10-13.

[6] 夏丽芳，张方玉，王仲朗. 云南省楚雄市茶花古树录 [M]. 昆明：云南科技出版社，2011.

[7] 叶少萍，张俊涛，曹芳怡，等. 立地环境改造对古树根系分布特征的影响——以44011111322000296号朴树为例 [J]. 林业与环境科学，2021，37(3)：75-80.

[8] 游慕贤. 中国古茶树 [J]. 国土绿化，2002(2)：28.

[9] 游慕贤. 中国茶花古树觅踪十年 [M]. 杭州：浙江科学技术出版社，2010.

[10] 张国豪，蔡孔瑜，田艳，等. 古树根境土壤改良及复壮效果评价——以重庆市铜梁区黄葛古树为例 [J]. 安徽农业科学，2021，49(12)：103-106，111.

[11] 张乐初，游慕贤. 我国野生山茶和山茶古树考察侧记 [J]. 花木盆景（花卉园艺），2001(3)：8-9，59.

[12] 张琴，张东方，孙成忠，等. 气候特征与药用植物地理分布的数值分析——以四川省为例 [J]. 中国现代中药，2018，20(2)：145-151.

[13] 周利. 川茶花品种图鉴 [M]. 重庆：重庆出版社，2011.

[14] 周利，李玲莉，宋春艳，等. 重庆地区川山茶古树资源调查及保护现状分析 [A]. 2021年中国植物园学术年会论文集 [C]，2021：80-84.

[15] 周利，李玲莉，宋春艳，等. 川渝地区川山茶古树树龄与地径关系研究 [J]. 湖北农业科学，2021，60(S2)：320-322.

References

[1] Gao J Y, PARKS C R, Du Y Q. Collected species of the genus camellia an illustrated outline[M]. Hangzhou: Zhejiang Science and Technology Press, 2005.

[2] Guan K Y, Li J Y, Wang Z L.Camellia of China[M]. Hangzhou: Zhejiang Science and Technology Press, 2014.

[3] LI J Y, NI S, WU H, *et al.* A rapid non-destructive tool to measure ages of living camellias[A]. Proceedings of Conghua International Historic Camellia Conservation Meeting[C], 2019: 23-30.

[4] LY/T2814-2017, Technical Regulation for Camellia japonica Production[S].

[5] Wang Y X, Gao L S, Zhao X H.Tree-ring width chronology of populus tremula and its relationship with the weather in Changbai mountain of northeast China[J]. Journal of Northeast Forestry University, 2013, 41(1):10-13.

[6] Xia L F, Zhang F Y, Wang Z L. Records of camellia ancient trees in Chuxiong city of Yunnan[M]. Kunming : Yunnan Science and Technology Press, 2011.

[7] Ye S Y, Zhang J T, Cao F Y, et al. Effects of site environment reconstruction on root distribution characteristics of ancient trees: the case of the No. 44011111322000296 celtis sinensis[J]. Forestry and Environmental Science, 2021, 37(3): 75-80.

[8] You M X. Ancient camellia trees in China[J]. Land Greening, 2002(2): 28.

[9] You M X. Tracking down Chinese ancient camellia trees for ten years[M]. Hangzhou: Zhejiang Science and Technology Press, 2010.

[10] Zhang G H, Cai K Y, Tian Y, et al. The evaluation on soil improvement and rejuvenation effect of the ancient trees-taking ancient ficus virens in tongliang district of Chongqing as an example[J]Journal of Anhui Agricultural Sciences, 2021, 49(12): 103-106, 111.

[11] Zhang L C, You M X. A sidelights on the investigation of wild Camellia and ancient camellia trees in China[J]. Flower Plant & Penjing, 2001(3): 8-10.

[12] Zhang Q, Zhang D F, Sun C Z, et al. Numerical analysis of climatic characteristics and geographical distribution of medicinal plants-taking Sichuan province as an example[J]. Modern Chinese Medicine, 2018,

20(2):145-151.

[13] Zhou L. Illustrated handbook for Sichuan camellias[M]. Chongqing: Chongqing Publishing House, 2011.

[14] Zhou L, Li L L, Song C Y, et al. Investigation and protection of ancient camellia tree resources in Chongqing area[A]. Proceedings of Botanical Society of China Congress [C], 2021: 43-48.

[15] Zhou L, Li L L, Song C Y, et al. Study on the relationship between age and ground diameter of ancient trees of Sichuan camellia in Sichuan and Chongqing areas[J].Hubei Agricultural Sciences, 2021, 60(S2): 320-322.

图书在版编目（CIP）数据

川山茶古树名录 / 周利主编. — 重庆：重庆出版社，2022.12
ISBN 978-7-229-17312-8

Ⅰ.①川… Ⅱ.①周… Ⅲ.①茶树－四川－名录 Ⅳ.①S571.1-62

中国版本图书馆CIP数据核字(2022)第252170号

川山茶古树名录
CHUAN SHANCHA GU SHU MINGLU

周　利　主编

责任编辑：王　娟　苏　杭
责任校对：朱彦谚
美术编辑：梁又双
装帧设计：梁又双　肖　琴

重庆出版集团
重庆出版社 出版

重庆市南岸区南滨路 162 号 1 幢　邮政编码：400061　http://www.cqph.com
重庆高迪彩色印刷有限公司印刷
重庆出版集团图书发行有限公司发行
E-MAIL:fxchu@cqph.com　邮购电话：023-61520656
全国新华书店经销

开本：889mm×1194mm　1/16　印张：22.75　字数：380千
2022年12月第1版　2022年12月第1次印刷
ISBN 978-7-229-17312-8
定价：328.00元

如有印装质量问题，请向本集团图书发行有限公司调换：023-61520678

版权所有　侵权必究